A Scientific Adventure

Reflections on the Riddle of Relativity

Ian McCausland

Apeiron
Montreal

Published by C. Roy Keys Inc.
4405, rue St-Dominique
Montreal, Quebec H2W 2B2 Canada
http://redshift.vif.com

First Published 2011

Library and Archives Canada Cataloguing in Publication

McCausland, Ian, 1929-
A scientific adventure : reflections on the riddle of relativity / Ian McCausland.

Includes bibliographical references.
ISBN 978-0-9864926-6-2

1. Relativity (Physics). 2. Dingle, Herbert, 1890-. 3. Good, Irving John. 4. Einstein, Albert, 1879-1955. I. Title.

QC173.585.M32 2011 530.11 C2011-907504-0

Cover Image: M. C. Escher's "Relativity" in *LEGO®*, reproduced by permission of Andrew Lipson© (www.andrewlipson.com).

Table of Contents

Preface

> True philosophers, who are only eager for truth and knowledge, never regard themselves as already so thoroughly informed, but that they welcome further information from whomsoever and from wheresoever it may come; nor are they so narrow-minded as to imagine any of the arts or sciences transmitted to us by the ancients, in such a state of forwardness or completeness, that nothing is left for the ingenuity and industry of others.
>
> William Harvey, *An Anatomical Dissertation upon the Movement of the Heart and Blood in Animals*, 1628.

This book is the story of my collaboration with Professor Herbert Dingle (1890-1978) during the last six years of his life, and of my attempts to continue some of his work after his death. Herbert Dingle was an eminent English scientist and was Professor of the History and Philosophy of Science at University College, London. My collaboration with him was in connection with his claim that there is an internal inconsistency in one of the most revered theories of science—Albert Einstein's special relativity theory. Professor Dingle described in his book *Science at the Crossroads* (Dingle, 1972) his attempts to persuade the scientific community that there is an inconsistency in the theory. Although he did not succeed in convincing the scientific community of the inconsistency, the story did not end there, and this book is an attempt to provide a fair appraisal of his thesis and the way the scientific community reacted to it.

In spite of their belief that Dingle's argument was wrong, some writers have expressed their belief that Dingle was badly treated. One of them (Grant, 2006) wrote as follows:

> Dingle was shabbily treated by the scientific community, which seemed to feel that if it ignored him for long enough he would simply go away. His letters on the subject were refused publication in *Nature*, and there seems to have been a degree of petty academic skullduggery at work which reflects little credit on our physicists.

Since special relativity is so highly revered, most scientists are so strongly convinced of the soundness of the theory that they are unwilling even to consider the possibility that Dingle might have been right, yet in many cases their reasons for believing him wrong are not valid. In this book I discuss some of the invalid arguments that have been used to try to refute Dingle's thesis, and also present evidence of some of the petty skullduggery that he experienced.

Dingle based his study of the special theory on Einstein's original paper on the subject, published in 1905 in the German journal *Annalen der Physik* under the title "Zur Elektrodynamik bewegter Körper," and on the widely accepted English translation which appeared under the title "On the Electrodynamics of Moving Bodies" in a collection of papers (Lorentz *et al.*, 1923) which is still widely available as a Dover paperback. The question that Dingle asked about the special theory is essentially this: Is there, or is there not, an internal inconsistency in Einstein's special theory *as described in his original paper on the subject?*

The answer to that question depends on the logical properties of the theory, as expounded in Einstein's original paper; so it can be answered only by examining that paper or a reliable translation of it. The theory is a product of a human mind, so the answer to the question of its self-consistency does not depend on the physical properties of the rest of the world and cannot be obtained by doing physical experiments. Furthermore, if the question had been asked in 1905, just after the original paper had been published, it could have been answered definitively at that time by examining the paper; so the answer cannot depend on any subsequent events. Yet some of Dingle's opponents have used appeals to experimental evidence and to the general relativity theory, which did not appear until about a decade after the special theory, to try to refute his claim of an inconsistency.

I show in Chapter 8 that, if a theory has an internal inconsistency, it is possible to derive from the theory, using valid rules of logic, any proposition whatever. So it is not possible to refute a claim of an inconsistency by appealing to experimental evidence, since an inconsistent theory could match all the experimental evidence that is accepted as supporting the theory. Since the question of an inconsistency in the theory is a logical problem and not a mathematical problem, I have avoided mathematical equations in this book; the only equation in the book is in a quotation in Appendix I from one of Dingle's arguments.

I review some of the published debate, and point out many faults in the arguments of the defenders of the special theory. Although I am not a physicist or a mathematician, I am an electrical engineer, and I believe that I am well qualified to write about the controversy that Herbert Dingle aroused by his claim of an inconsistency in the special theory. I believe that I am also well qualified to write about Herbert Dingle's role in the controversy, since I was in frequent correspondence with him for several years before his death and, as mentioned below, he also sent me copies of some of his correspondence with other people.

I have already told some of the story of my collaboration with Herbert Dingle in a book called *The Relativity Question* (McCausland, 1988). One of the purposes of that book was to put on record a fairly large body of correspondence between Professor Dingle and various influential people, including the President of the Royal Society, the Minister of State for Science and Education, the Archbishop of Canterbury, Members of the Council for Science and Society, and the Editors of *Nature*. Copies of that correspondence had been sent to me by Dingle or by another of his collaborators, Mr. Mark Haymon, with a view to eventual publication. Since that body of correspondence is now on record there is no need to publish it again; it does not appear in the present essay except for some minor comments. Copies of some of Dingle's other correspondence that appears in the present book were acquired in the course of visits to the archives of Imperial College, London.

One of the important parts of this book is the story of my long debate with the distinguished mathematician Professor I. J. Good, which was published in the pages of the journal *Physics Essays*. Although this had to end with Good's death in April 2009, he had already conceded a crucial error in his own argument, and my original argument that started that debate is still unrefuted and needs to be answered.

In addition to the story of my collaboration with Dingle, this book describes some of my explorations into other aspects of relativity and what people have written about it.

My title was suggested by the title of one of Herbert Dingle's books, *The Scientific Adventure* (Dingle, 1952). I am proud that I shared a small part of Herbert Dingle's scientific adventure., and I dedicate this book to his memory.

In the earlier stages of my adventure I received welcome support and encouragement from Mark Haymon, Herbert Dingle's lawyer and collaborator, and John Green, one of his former students. Later I had the good fortune to became acquainted by correspondence with Tom Phipps, one of the most vigorous critics of special relativity, and I was encouraged by him to publish a criticism of the special theory in *Physics Essays*. I was also fortunate in meeting Henry Bauer, who introduced me to Jack Good and helped to start our discussion. I wish to pay tribute to Jack Good's willingness to engage in a published debate in the pages of *Physics Essays*, and to Emilio Panarella, Editor of that journal, for his patience and cooperation in dealing with our papers. I am also grateful to Nick Percival and Ron Hatch for discussions relating to that debate. Other friends and colleagues who provided information, advice and moral support were Fred Doran, Peter Hayes, Vít Klemeš, Paul Marmet, Henry Palka, Harry Ricker and Richard Waldron.

I thank the Natural Philosophy Alliance for the opportunity to present several of my earlier publications, including *The Relativity Question*, on the World Science Database; special thanks are due to David Scott de Hilster and Greg Volk, This internet database is a useful public service to those who are interested in subjects that are rarely discussed in mainstream scientific journals.

I am grateful to the staff of the Archives of Imperial College, London, for providing access to Herbert Dingle's papers, and to the staff of the University of Toronto Library System for help in locating other materials.

The presentation of my story has been enormously improved by many very helpful suggestions of Roy Keys and David Roscoe. Very special thanks are due to Roy for publishing the book. It has been a great pleasure to work with him, and it is an honour to have the book published in his Apeiron series.

Special gratitude is due to my wife Joan for her quiet support of my activities in this part of our even longer adventure together.

Chapter 1

Introduction

> All scientific knowledge is provisional. Everything that science "knows," even the most mundane facts and long-established theories, is subject to reexamination as new information comes in. The latest ideas and data are the most provisional of all. Some recantations will be unavoidable. This is not a weakness of science; this is its glory. No endeavor rivals science in its incremental progress toward a more complete understanding of the observable world.
>
> *Scientific American* editorial, "In Science We Trust"
> December 2002, 14.

Albert Einstein (1879-1955) is one of the most famous scientists in history, and his fame rests mainly on his two theories of relativity, the Special Relativity Theory, which was set forth in a paper published in 1905, and the General Relativity Theory, published in final form in 1916. Einstein rose to great fame in November 1919 when it was announced that astronomical observations taken at a solar eclipse in May 1919 supported the General Theory. Ever since then he has been a prominent public figure, and his prestige and eminence have continued to grow with the passage of time.

The prestige of the two theories is enormous, both in the scientific community and in the wider population. Some writers have described the impact of the theories of relativity as if they had started a new era of civilization. For example, Judson (1980) described science as "our century's art," and referred to Einstein's paper on special relativity as if it were the herald of the 20th Century's art:

> The takeover can be dated more precisely than the beginning of most eras: Friday, June 30, 1905, will do, the day when Albert Einstein, a clerk in the Swiss patent office in Bern, submitted a thirty-one-page paper, "On the Electrodynamics of Moving Bodies," to the journal *Annalen der Physik*. No poem, no play, no piece of music written since then comes near the theory of relativity in its power, as one strains to apprehend it, to make the mind tremble with delight.

It seems to me that the date of *publication* of Einstein's paper would be more appropriate as the start of the era than the date of *submission*. Even so, the start of the era that was initiated by the general theory was even more precisely defined by Johnson (1991) on the first page of his book *A History of the Modern World*:

> The modern world began on 29 May 1919 when photographs of a solar eclipse, taken on the island of Principe off West Africa and at Sobral in Brazil, confirmed the truth of a new theory of the universe.

Since the total eclipse lasted about five minutes at each of the two sites at which the photographs were taken, Johnson's precision seems excessive, especially as it is so obviously inconsistent with the subtitle of his book, which is *From 1917 to the 1990s*. Again, it might be suggested that the more appropriate date for the start of the modern world would be the day on which the modern world found out about it, which was at the famous joint meeting of the Royal Society and the Royal Astronomical Society in London on 6 November 1919, at which the results obtained from the photographs taken at the solar eclipse were presented publicly.

My first exposure to relativity occurred in the second academic year (1947-48) of my undergraduate course in Electrical Engineering in The Queen's University of Belfast, Northern Ireland, when I was 18 years old. One of the requirements of one course was to give a short talk on a subject outside the curriculum, and one of my fellow students, Abdul Daood, gave a talk on relativity. The description of his talk that I now give is not meant to be a rigorous statement of the problem of relativity, but only the impression that I received at the time. He started by saying that absolute motion was meaningless; one can only deal with the motions of bodies relative to other bodies. In the course of his description of the special theory he described the clock paradox, according to which, if there were two synchronized clocks at the same place, and if one clock was moved away and returned at very high speed, then the clock that had been moved away and returned would no longer be synchronized with the other, but would show an earlier reading. The paradox arose because, if absolute motion is meaningless and only the relative motion of the clocks counts, either clock can be taken to have moved. I might not have been bothered by that paradox, were it not that the speaker went on to extend the same idea from a pair of clocks to a pair of twins, and stated that it is possible, according to the theory, for one of a pair of twins to travel in space for a long time at a very high speed and on his return to find himself younger than the twin who had stayed at home. Again, the twin paradox arose because it appeared that either twin could be taken as the moving one.

Although I immediately felt very sceptical about that description of the paradox, I did not know enough about the subject to feel confident about my scepticism. I was much too busy with my studies to spend time in studying the subject to the point of understanding the resolution of the paradox. As the years passed, I was aware that there was some debate on the subject, and that the main antagonists were Professor W. H. McCrea, who believed that the special theory entailed asymmetrical ageing, and Professor Herbert Dingle, who believed that it did not. The fact that both of these eminent professors had written books on relativity (McCrea, 1935; Dingle, 1940) seemed to make it obvious that there was little point in my trying to decide the matter for myself. I did read some popular books on relativity, as I describe in the next chapter, but they did not provide a solution to my problem.

Eventually, as a Professor of Electrical Engineering at the University of Toronto, I took sabbatical leave during the academic year 1971-72, and spent the year at Churchill College and the Department of Engineering at Cambridge University, from which I had obtained a Ph.D. degree in 1964. I soon discovered that there was a debate still going on about special relativity, and that Professors Dingle and McCrea were still involved. The particular chapter of the debate that I encountered started with an article by John Taylor, Professor of mathematics at King's College, London, in *The Listener* dated 11 November 1971 (Taylor, 1971b). In the article Taylor reported on the experiment that had recently been carried out by Joseph Hafele and Richard Keating, in which a very accurate clock was flown around the world and then compared with the reading of a similar clock that had stayed in Washington. According to Taylor, if the readings of the two clocks did not differ by the expected amount, special relativity would have been contradicted. He claimed that the experiment had been successful.

Professor Dingle replied to Taylor's article by a letter in the next issue of *The Listener*, dated 18 November 1971, and further letters were published from other participants until the 13 January 1972 issue, when the editor declared the correspondence closed.

One of the exchanges in the correspondence was very striking: in a letter in the 2 December issue, Dingle asked a question of Taylor in the following words:

> Professor Taylor and others now say in effect that the theory only requires that each clock appears to an observer with the other to go slow. I therefore ask him to answer, with a plain yes or no, the following question: does he consider that Einstein meant that when the clocks, originally at A and B, met at B, observers accompanying them throughout the process would disagree on which was giving the earlier reading?

In the 9 December issue Taylor published a reply but avoided giving such a plain answer. The evasion was so striking that the journalist Bernard Levin referred to it in an article "A plain man's guide to the theory that Einstein was wrong" in the 21 December issue of *The Times* (Levin, 1961). Levin quoted the question that Dingle had asked, and enough of Taylor's answer to show that Taylor had not answered the question. However, he did not mention what I thought was an even more striking feature of Taylor's reply to Dingle, how he ended his published letter in the following words:

> I am sure Professor Dingle doesn't wish to come under the latter heading in the proverb: 'Those that can, create; those that can't, criticise.'

Professor Taylor's remark was my first experience of the vilification of Herbert Dingle that has been a prominent feature of the reaction to his criticisms. The inappropriateness of Taylor's insult may be judged by the fact that, at that time, Professor Dingle's published writings on relativity had spanned a period of almost fifty years, and he wrote his first book on the subject several years before Professor Taylor was born in 1931. It is well known that a common way of trying to hide the weakness of one's own argument is to insult one's opponent, and Taylor's insult is an excellent example. I have since become aware of even more vicious insults to Dingle in anonymous comments on the internet, but I shall not provide instructions for finding these.

It is interesting also to compare Taylor's comment with what he himself had written in the same magazine only two months earlier, (Taylor, 1971a):

> I felt very strongly that science is too scientific to be left to the scientists. They are often swayed too strongly by their emotions to take a properly detached view, and can cause untold harm to the future development of science.

It is also interesting to consider the last paragraph of Professor Taylor's article in *The Listener* dated 11 November 1971 (Taylor, 1971b), which refers to the Hafele-Keating experiment as follows:

> The experiment has worked. It didn't really need doing, since Einstein's theory had already been tested under far more extreme conditions. But such a test had to be performed, if only to lay the doubting Thomases to rest. *Requiescant in pace.*

It seems strange that a scientist should state that an experiment "didn't really need doing," implying that its result could have been known (rather than merely predicted) in advance. An experiment, by definition, carries no guarantee of any particular outcome. Taylor's statement is, in my opinion, completely unscientific, but is reasonably typical of the complacent certainty of their own rightness which is a feature of the attitude of so many relativists. In any case, as I show in Chapters 8 and 11, a claim of an inconsistency cannot be refuted by experiment.

When I realized that the problems of relativity were still being debated, it seemed to me that, if I was ever to satisfy myself about the truth of the matter, I would never have a better opportunity again. My decision to study the problem more thoroughly marked the start of the main part of my scientific adventure.

At first I did not even realize that the subject of the debate had changed. Professor Dingle was no longer claiming merely that the theory did not entail asymmetrical ageing of space travellers; he was claiming that the theory was self-contradictory. Unfortunately some of Dingle's scientific opponents were also unaware of the change. For example, a letter appeared in *The Times* on 8 January 1972, written by Professor R. A. Lyttleton, F.R.S., as a reply to Bernard

Levin's article, in which the following paragraph shows that he did not realize that the subject of discussion had changed:

> In brief, what Dingle has steadfastly maintained these many years against all comers is this: That if Peter and Paul are identical twins, and Paul goes on a journey leaving Peter to stay at home, then when Paul returns he will still be exactly the same age as his brother.

Dingle replied in a letter that was published in *The Times* on 26 January, pointing out that Lyttleton had got everything wrong—even the point at issue. He went on to describe his change of view as follows:

> Regarding the immeasurably less important clock paradox, Lyttleton is again wrong in saying that I have denied asymmetrical ageing for many years. Fifteen years ago, when I believed special relativity true, I indeed thought it impossible, but I soon discovered my error, and for more than 13 years have held the question open... Despite the mu-mesons and their kind, I think asymmetrical ageing extremely unlikely, but that is an opinion; the falsity of the special relativity theory (not necessarily of the relativity of motion) I regard as proved.

There was no subsequent letter in *The Times* from Lyttleton acknowledging his error.

Dingle later published both the above-mentioned letters from *The Times* on pages 10-12 of his book *Science at the Crossroads* (Dingle, 1972), to show that it is important to distinguish between the earlier debate on the clock or twin paradox and the later debate on the consistency or inconsistency of the special theory.

In early 1972, while I was still in Cambridge, I wrote to Professor Dingle to discuss some of my ideas on the problem, and we exchanged a few letters. I had hoped that I might have been able to meet him while I was living in England at the time, but no suitable opportunity arose. However, I wrote to him again after the publication of *Science at the Crossroads*, by which time I was back in Toronto, and this started a correspondence that continued uninterrupted until his death in 1978. After his death I continued my activity in studying the special theory and writing about it, as described in the rest of this book.

PART ONE

Surveying the Literature

In this part I describe some of my survey of the literature about relativity in an attempt to resolve the clock paradox or twin paradox, at least to my own satisfaction. In both the chapters of this part, I describe both the publications that I studied at the time I was still learning and my later investigations into various aspects of the subject, and I include discussion of some of the unfair representations of Herbert Dingle's arguments made by various people. Most of the survey in Chapter 2 was done before I went to Cambridge on sabbatical leave, and the survey in Chapter 3 began during my stay in Cambridge, after I discovered that the debate on the clock problem was still continuing.

Chapter 2

My Early Studies of Relativity

> Myself when young did eagerly frequent,
> Doctor and Saint, and heard great argument
> About it and about: yet evermore
> Came out by the same door as in I went.
>
> Edward Fitzgerald, The Rubaiyat of Omar Khayyam

During the years after my first scepticism about the twin paradox and before becoming active in the controversy in 1971, I read some popular presentations of relativity. The first one I read was Lincoln Barnett's book *The Universe and Dr. Einstein* (Barnett, 1949), which had a Foreword by Albert Einstein. It was in its day an excellent book, but is now rather out of date. Unfortunately Barnett did not provide the answer to my problem. About the only thing he wrote about the twin paradox was the following (p. 51):

> In a Buck Rogers realm of fantasy, it is possible to imagine some future cosmic explorer boarding an atom-propelled space ship, ranging the void at 167,000 miles per second, and returning to earth after ten terrestrial years to find himself physically only five years older.

I also read James A. Coleman's book *Relativity for the Layman*. My reading of this book led to a very interesting study in itself, which continued well after I had become actively involved in the controversy in 1971. My interest was aroused when I discovered in the 1959 version of his book (Coleman, 1959) his argument that asymmetrical ageing does not occur. His attitude to asymmetrical ageing in that edition of the book was, if anything, scornful. For example, he wrote (p. 71):

> Not only has the paradox been seriously debated in scientific journals, but many people are claiming, in all seriousness, that people who continually engage in space travel will remain perpetually young. This is not true, as the following explanation of the apparent paradox will now show.

There followed an explanation stating that the special theory does not apply to the case considered, and if the general theory is used the slowing down during constant velocity is cancelled when the rocket's velocity changes during launching and landing; this led Coleman to the following conclusion: "Hence, there is no permanent effect and, of course, no paradox."

Not only did Coleman reject asymmetrical ageing, but this version of his book also carried an endorsement by Einstein himself, in the following quotation printed on the inside front cover of the book.

> 'Gives a really clear idea of the problem, especially the development of our knowledge concerning the propagation of light and the difficulties which arose from the apparently inevitable introduction of the ether. The relativity of motion judged from a cinematic point of view is vividly demonstrated and also the principle of equivalence.'

Naturally I thought that my scepticism about the twin paradox was justified, if Einstein himself had endorsed a book that said asymmetrical ageing did not happen. Unfortunately my feeling of justification did not last, as another version of the book appeared (Coleman, 1969) which reversed the author's earlier position on the twin paradox, and supported asymmetrical ageing. In his "Preface to Revised Edition" (p. 12) Coleman gave his reason for this reversal in the following words:

> In the case of the time, or clock, paradox the long-smouldering disagreement regarding its explanation finally erupted in 1957-59 in a voluminous outpouring of papers on the topic. I have taken advantage of the clarification of this topic which has resulted and have completely rewritten the section dealing with it to conform to the present views of the majority of those theoretical physicists who are deemed to be most knowledgeable about relativity.

Coleman did not say how he deemed those physicists to be most knowledgeable about relativity. In the rewritten section, he contradicted his earlier statement by saying that the special theory is sufficient to solve the problem, and again wrote that there is no paradox.

On the outside back cover of that 1969 version of the book, below the words "Second Edition" there appeared exactly the same quotation from Einstein that had appeared in the 1959 version. I was surprised to find out that Einstein appeared to have changed his mind and believed in asymmetrical ageing after all. Since he had died in 1955, it was not immediately clear how he had communicated his new insight to be incorporated in the 1969 version of the book, and eventually I was so overcome with curiosity that I wrote to the publisher, enclosing a photocopy of the quotation from the 1969 version, and asked for its source. My enquiry was referred to Professor Coleman, and I received a letter from him telling me that the quotation was from a letter from Einstein to him, dated February 7, 1955, which was only two months before Einstein's death.

I remained very puzzled, since it still appeared that Einstein had endorsed the earlier version of the book that presented the unorthodox solution of the twin paradox. It was not until I read a review of *Relativity for the Layman* (Moore, 1959) that I realized what had actually happened. Moore's review was of a Mentor book published by The New American Library in 1958, which is essentially the same as the 1959 Penguin/Pelican edition. After mentioning some small errors in the book, Moore went on to say "These errors are minor, however, when compared with the author's handling of the 'clock paradox'—an error that did not appear in the first edition, which made no attempt to answer the question." That was the first clue I had ever received that the 1959 edition was different from the original 1954 edition. Since the 1959 edition carried a preface dated 17 March 1954, it had seemed reasonable to assume that the 1954 and 1959 editions were the same, but when I finally found the first edition published by the William-Frederick Press in 1954 I found that they were different, even though their prefaces were identical, word for word.

In the first edition, Coleman (1954) mentioned the paradox but did not take a position on its resolution. Towards the end of his Chapter 3, on "The Special Theory of Relativity," he described a hypothetical journey by rocket men to a distant star, and mentioned two potential outcomes: one was that the rocket men would age less than their stay-at-home relatives, the other was that they would age more. After the description of these different outcomes, he ended the chapter with the following two paragraphs:

> This is the paradox: At the end of such a rocket trip will the people on earth be older than the rocket men, or will the rocket men be older than the people on earth? Both views are correct according to the Special Theory. Yet they are contradictory and both cannot be true.
>
> You are now left to ponder this situation in quiet contemplation, without hindrance from the author.

This obviously cleared up part of my problem. I realized that what Einstein wrote about the book was based on the first edition (Coleman, 1954), which did not take a position on the paradox. It may be noted here that, if Coleman had put together his two statements that "both views are correct according to the Special Theory" and "both cannot be true" he could have arrived at the conclusion that Herbert Dingle reached, that the special theory in self-contradictory because both the contradictory results are correct according to the Special Theory.

It is clear that the 1959 Penguin/Pelican edition, which is explicitly described on the cover as Second Edition, is in fact the third version of the book. Even then, the above misleading collection of information about the various editions of the book was not the only surprise I received. Eventually I became curious as to what Einstein had actually written about the book, and in 2006 I acquired from the Einstein archives in Jerusalem a copy of Einstein's letter to Coleman. I was surprised to find that the second sentence of the Einstein quotation that appeared in the 1959 and 1969 Pelican/Penguin versions of the book (The relativity of motion judged from a cinematic point of view is vividly demonstrated and also the principle of equivalence.) did not appear in the letter, either the English version of the letter or the German draft in Einstein's handwriting.

I later discovered that Coleman had written an autobiography (Coleman 1996), and I acquired a copy to try to find out how the mysterious sentence had been added. Although the book about relativity and the correspondence with Einstein are mentioned (p.73), there is no information about the added sentence.

Needless to say, Coleman's intellectual gymnastics on the subject of the twin paradox did not help me much in trying to come to my own solution of the problem.

In the course of my studies of relativity I also read books such as Sir James Jeans's *The Mysterious Universe* (Jeans, 1930). I found that he was very respectful toward mathematicians but much less respectful to engineers. For example, he wrote (p. 86), referring to Einstein's 1905 paper on special relativity: "with its publication, the study of the inner workings of nature passed from the engineer-scientist to the mathematician." He went on to say (p. 111-2) that "no one except a mathematician need ever hope fully to understand those branches of science which try to unravel the fundamental nature of the universe—the theory of relativity, the theory of quanta and the wave-mechanics." He also rejected the possibility that the universe might have been planned by a biologist or an engineer, and wrote (p. 122) that "from the intrinsic evidence of his creation, the Great Architect of the Universe now begins to appear as a pure mathematician." Jeans was, of course, a mathematician.

Although Jeans seemed a little scornful of engineers, another eminent scientist (Frisch, 1979) stated that the special theory was part of engineering, and one of the grandiose titles accorded to Einstein during the centennial of his miraculous year in 2005 appeared in three volumes published by Wiley with the title *Albert Einstein: Chief Engineer of the Universe.*

When Jeans wrote that the Great Architect of the Universe appeared from the evidence of his creation to be a pure mathematician, he was clearly showing a belief in an intelligent male designer. Did anyone point out that such a belief was outside the boundaries of science? Yes, Herbert Dingle did. In his book *Science and Human Experience* (Dingle, 1931, 131) he wrote that much interest had been aroused by Jeans's view that "the Great Architect of the Universe now begins to appear as a pure mathematician," and went on to say:

> Sir James is himself an eminent mathematician, and his statements are widely accepted as the embodiment of current scientific thought. In some respects, of course, they are—as, for instance, in his fascinating exposition of the theory of relativity—but it cannot be too clearly understood that when he makes the remark just quoted, and others relating thereto, he is speaking for himself alone, and is expressing not a scientific deduction but a metaphysical one, requiring for its validity an assumption for which Science gives no warrant.

In spite of Jeans's comments, the mathematical knowledge required for the understanding of the behaviour of clocks in motion in special relativity is not very sophisticated. One of my early published papers on relativity (McCausland, 1974) showed an analogy between the equations of the Lorentz transformation, which is the mathematical core of the special theory, and the transmission line equations used by electrical engineers.

Another claim to the importance of mathematics in relativity was made by Leonard B. Loeb and Arthur S. Adams, who wrote as follows (Loeb and Adams, 1933, 167):

> Practically all the men who attack relativity are either fanatics or men who are so poorly equipped mathematically that they are incapable of understanding or following the processes involved. The difficulties in the understanding and interpretation of relativity result from the nature of the mathematics which Einstein uses to develop his theory.

It is interesting to see that defenders of relativity, as early as 1933, were already using the strategy of insulting the critics by describing them as attackers and suggesting that they are either fanatics or mathematical ignoramuses. Another interesting feature of the same book is the authors' own ignorance of the history of relativity, as shown by the following quotation from page 165 of their book:

> It was not until 1919 that President Campbell of the University of California and members of his expedition were able to make a successful and accurate set of observations of the positions of the stars in the neighbourhood of the sun during an eclipse... The results of the measurements made by President Campbell and his associates completely verified the Einstein theory.

This is a complete misrepresentation of the history of the 1919 eclipse expeditions, in which Eddington and British colleagues made the observations that were claimed to support Einstein's general theory.

Chapter 3

The Space-Travellers

There was a young lady named Bright
Whose speed was far faster than light;
She set out one day
In a relative way
And returned on the previous night.

A. H. R. Buller, *Punch*, 1923.

When I became aware in 1971 that a relativity controversy was going on, and decided to study the problem more thoroughly, I first thought that I should try to resolve the twin paradox. However, when I first wrote to Herbert Dingle he made it clear that he had gone beyond that paradox and was not interested in discussing possible resolutions of it. This is part of what he wrote in his first letter to me, in December 1971:

> A paradox arises when two apparently sound but contradictory conclusions, X and Y, result from the same premises P. It can be resolved only by (1) disproving X; (2) disproving Y; or (3) finding a contradiction inherent in P. All additional proofs of X or Y do nothing without a disproof of Y or X. Years ago I believed P sound and X (symmetrical ageing) sound, and tried to disprove Y. I could not do so to my full satisfaction, and that led me to discover a contradiction in P. ... It therefore seems to me unimportant to compare the merits of X and Y.

In another letter, in January 1972, he mentioned the book *Time and the Space Traveller* (Marder, 1971) which had recently been published, and suggested that I might find it useful. I quickly acquired a copy of the book, and found it a useful source of historical information on the clock problem. For example it has 30 references to publications by Dingle. Although some of those references refer to Dingle's claim that there is an inconsistency in the special theory, the book does not deal with these in detail but concentrates on the debate on the clock or twin paradox. In a similar way, I shall deal only with the literature on the clock paradox itself in the present chapter.

Let us start with two examples, both of which were cited by Marder in his book. The first was written by Professor W. H. McCrea (1957):

> Were a new Isaac Newton born to-day, we could send him space-travelling so as to return to us in thirty years time at, say, the age of three. This would be in accordance with the theory of relativity and with experimental tests of the theory—and all we should get for our efforts would be a retarded child.

The last phrase might make it appear that McCrea was joking, but the context makes it clear that he was serious. *Nature* is not known for its sense of humour, and I suspect that a 21st-Century editor would have deleted that phrase because it is not politically correct to describe a child as being retarded.

The second example appeared in an article in *The Times* (London) on September 19, 1956, from a Special Correspondent in Rome, under a prominent heading "NEAR-IMMORTALITY" IN SPACE TRAVEL.

> A forecast on the distant future of space flight was made here to-day by Professor G. A. Crocco, president of the Italian rocket society. He told the international astro-nautical congress that when the era of inter-stellar travel, at close to the speed of light, was reached Einstein's "time expansion" phenomenon would come into play, and while centuries passed by travellers would experience only the passage of minutes and would become "almost immortal." But he added a warning that as soon as the ship's fuel gave out this near-immortality would cease.

Suppose that these two statements, McCrea's and Crocco's, are both scientifically justified, and let us examine some possible consequences. Suppose, for example, that we have a pair of twins, born in the year 2000, and that in 2020, at age twenty, one of them goes space-travelling at the speed used in McCrea's example, to return 600 years later (in 2620) at the age of eighty. Suppose that when the traveller returns he finds that his twin died at the age of 80 in 2080, 540 years earlier, and that he also dies at the age of 80, just after his return to earth. The fact that he died in 2620 would not mean that he had lived 620 years; he and his twin brother would both have died at the age of 80. Although the traveller's tombstone would look very impressive with the dates 2000-2620, he would have been no closer to immortality than his stay-at-home brother, who lived a full and happy life on earth while the traveller spent the last 60 years of his life in a metal can eating spacecraft food and drinking his own recycled urine.

In *A Threefold Cord* (Samuel and Dingle, 1961, 79), Dingle quoted the above statement by Professor McCrea (without including the phrase about the retarded child), and went on to comment on this by saying that "Notwithstanding its stark inaccuracy—no such experimental tests have ever been made—this nonsense was published in the leading scientific journal." Later (p. 81) he remarked that "the public has learned to trust scientists' assertions, no matter how extravagant they may appear" and went on to say:

> Asymmetrical ageing is not in itself more astonishing than television or photographing the far side of the Moon: why should not the ordinary person believe that he can postpone the date of death indefinitely, in view of the authoritative pronouncements I have mentioned? And what may not be the social consequences of such a belief?

Dingle's wording is significant, the space traveller does not live longer—he merely postpones the date of his death—but the statement by Crocco quoted from *The Times* certainly gives the impression that people could live longer.

Dingle also commented on the above-mentioned statement in *The Times* by writing as follows in another article (Dingle, 1961b):

> Now that we are on the verge of attaining high speeds in space travel it is no trifle that people should be led to believe that they can postpone death in this way, and it needs little imagination to foresee the evil consequences that might ensue if this belief is acted upon and found illusory.

Also, in his first article dissenting from the orthodox view of the twin paradox in *Nature*, Dingle (1956) had written as follows:

> It needs little imagination to see that if the public is led to believe that there is scientific sanction for the idea that it is *possible* to postpone the date of one's death by space travel, some very undesirable consequences might ensue.

As I continued to study the history of this controversy in later years, I discovered that some of the reports on these discussions have been quite unfair to Dingle. Consider, for example, Stanley Goldberg's book *Understanding Relativity* (Goldberg, 1984). On page 161 he described some of the discussion between Dingle and McCrea during the 1950s, and made two statements that make it appear that Dingle was highly unscientific in his arguments. For example, he wrote that, in answer to McCrea's claim that asymmetrical ageing would actually occur, "Dingle countered that it was improper to think that one could seriously propose this as a way of extending one's lifetime, therefore, it could not possibly happen." Goldberg ended his account of the discussion between Dingle and McCrea by saying "When Dingle objected that it was immoral to postpone the date of death, the argument lapsed."

When I first read Goldberg's book I was not as familiar with the literature on the subject as I later became, but, although I did not know what Dingle had written on the subject, I did not think it likely that he would have used such unscientific arguments. Since Goldberg did not give any specific reference by which I could check his statement, I wrote to him in December 1993 to ask for the exact reference to Dingle's objection that it was immoral to postpone the date of death, but I did not receive any reply. I assume that what he was referring to was one or other of the three Dingle quotations above. But Dingle did not say that it was immoral to postpone the date of death; he said it was wrong to encourage people to believe that it was possible to do so.

In 2008 I came across an article that had appeared in the April 1998 issue of the *Journal of the Royal Astronomical Society of Canada*, in which an editor's note appended to an article on the twin paradox (Sarty, 1998) referred to the discussion between Dingle and McCrea and stated: "The argument finally lapsed when Dingle was reduced to objecting to relativity on the grounds that it would be immoral to postpone the date of one's death by relativistic means!" I was struck by the similarity between that wording and Goldberg's, with the added twist that he *was reduced to* that particular argument. Although I did not feel optimistic of any better result than my letter to Goldberg had achieved, I wrote to the person who had been editor when the item had been published. He had not written it himself, but gave me the name of the person who had written it: Professor Doug Forbes, a professor of astronomy and physics at Sir Wilfred Grenfell College in Corner Brook, Newfoundland. After some email discussion with Professor Forbes, including my sending him some notes I had written about Goldberg's remarks (along the lines of the discussion shown above), he admitted that he had probably used Goldberg as a secondary source. In an email dated 17 September 2008 he offered, without any prompting on my part, to write to the present editor to clear the matter up. More than a year later, on 3 October 2009, I sent him an email asking whether he had been able to clear the matter up as he had suggested. Although I received no reply, I think I have established that Dingle did not use the illogical arguments he was accused of.

In any case, the problem of acceleration rules out the possibility of attaining relativistic speeds unless enormously large distances are considered, such as those described by Marder (p. 21) in which an acceleration equal to g (the acceleration of gravity at the earth's surface) is maintained for a period of twelve astronaut years, or forty-two terrestrial years. Hermann Bondi, in his book *Relativity and Common Sense* (1964), also provided an interesting assessment of the limitations imposed by acceleration. He asked the question "How far can we travel in space, subject to our biological limitation?" He considered two cases (p. 153). One involved the acceleration of gravity for a total period of 40 years, in which time we would reach 24,000 light-years from Earth and return "to an Earth 48,004 years older than when we left it." The other case involves 2g, twice the acceleration of gravity, and this is what he said about that:

> If we are capable of taking 2g for forty years then we could travel to distant galaxies over 600 million light-years away, and would return correspondingly to an Earth over 1200 million years older. That we come back only to tell a much later generation is a serious matter, as is the fact that our most advanced rocket engineers could not dream of producing a rocket capable of maintaining such accelerations for the periods in question. This limitation, however, is a matter of technology, not of biology.

On the contrary, living in a spaceship with an acceleration of 2g for forty years would be far beyond the physiological and psychological limits of human endurance. Speculations like Bondi's are not science, or even science fiction; they are simply absurd.

Although he did consider the acceleration problem in part of his book, Marder allowed his imagination free rein in his chapter on "The Doubters." For example he wrote (p. 140), to show how time-dilation could "in principle" be used to advantage, "An individual could, for example, take a short vacation in space while his laboratory computer toiled for years on his calculations." It is difficult to believe that any scientist would be foolish enough to consider embarking on such an absurd expedition. After being absent from his laboratory for years, receiving no income and publishing no papers, the scientist's knowledge would be years out of date, his wife would be years older, his computer would be obsolete, and someone else would probably have already done the same calculations on a faster computer and published the results.

In any case such an expedition would be completely impossible, since there is no humanly survivable flight plan that would allow it to be accomplished. Suppose, for example, that our hero accelerated at 20 times the acceleration of gravity for a week, or 10 times that acceleration for two weeks, depending on the duration of the "short vacation." Even if we imagine that he could survive such an ordeal, his speed at the end of the acceleration would be less than 40 percent of the speed of light, and at that speed the change in the rate of passage of time would be less than 10 percent.

Another brilliant suggestion by Marder (p. 140) is that incurably ill people might go on space vacations "to await the discovery of a cure on earth." He was not joking: "This seems far-fetched, but it *might* one day be possible, and could well be preferable to the artificial hibernation of the incurably ill (by deep-freezing) as some have seriously suggested." [Italics in the original.] Anyone who seriously suggests that it might one day be possible to send incurably ill patients into space until a cure for their disease had been found on earth should make a realistic estimate of the mass of a spaceship that would be required to hold such a patient, with the necessary caregivers and the supplies of food and fuel that would be required, and the energy required to accelerate all of that to a high enough speed to make a significant difference in the rate of passage of time (and later to decelerate it), not to mention the difficulty of finding people willing to act as caregivers for the space-travelling patients. I would grant that it could well be preferable to the artificial hibernation of the incurably ill by deep-freezing, though it is my understanding that the cryonics people wait until their clients are actually dead before freezing them. It could well be preferable to the freezing, even if the spacecraft never left the ground.

As an epigraph to his chapter on "The Doubters," Marder used the following quotation from Eddington's book *The Mathematical Theory of Relativity*: "There are no doubt minds which have not this predisposition to regard as substantial the things which are permanent; but we shut them up in lunatic asylums." If those who doubt the pronouncements of Marder and like-minded people are to be shut up in lunatic asylums, what should be done with the people who accept those pronouncements uncritically? Physicists should not pretend to be qualified as sanity inspectors.

Robert L. Park (2008, 209) gave some estimates of the energy required for space travel. Park discussed with his university class the possibility of visiting the nearest star, making realistic assumptions about the number of people in the crew, the size of the spacecraft, and the time required for the trip. This is what they found:

> The energy needed for a trip to the nearest star in the lifetime of a human... is many thousands of times greater than all the energy that is expended on Earth for a year. So great, in fact, that there is no point in quibbling over the assumptions. It's just out of the question.

We might compare Park's assessment with one published by Dyson in 1968, as quoted by Marder (p. 110):

> Assuming a long-term expansion of the U.S. Gross National Product at the rate of 4 per cent per annum, in order to meet the cost, Dyson predicts that 'about 200 years from now, barring a catastrophe, the first interstellar voyages will begin'.

Dyson's paper (1968) was "based on a lecture given at the Belfer Graduate School of Science, Yeshiva University, in January 1968 as an entertainment between semesters." It is not entirely clear how seriously his estimate was meant to be taken, but I think the state of the U.S. economy more than forty years later might prompt a revision of the estimate.

How much could an astronaut *really* postpone the date of death by space travel? Marder (p. 97) wrote that "those who take only the four-day excursion to Mars save a mere half second." A return flight from Earth to Mars that could feasibly be undertaken by a human, which would last for months, would cause an even shorter time difference. If it were possible to travel at almost the speed of light from Earth to Mars and back, at their furthest separation, the time difference would be less than an hour. Personally, I think it extremely unlikely that humans will ever make a return trip to Mars, and I think it even more unlikely that it would be worth the cost. An interesting assessment of the problems that would have to be overcome, with intimate details of the effects of space travel on the organs of the human body, is described by Mary Roach in her book *Packing for Mars* (2010).

The scientific study of time travel continues in the 21st Century, going beyond the predictions of special relativity, and we now have what is called the grandfather paradox. Some science writers have discussed, in apparent seriousness, whether the laws of physics would allow a man to go far enough backwards in time to be able to kill his own grandfather before he had performed the function necessary to becoming a grandfather. If the traveller were to continue his super-Oedipal activities by marrying his grandmother, he could generate a whole sequence of paradoxes.

Although my survey of the literature on the twin paradox provided me with a good knowledge of the background of the problem, I needed to come to grips with the problem of a possible inconsistency in the special theory, and that became possible with the publication of *Science at the Crossroads* in 1972.

PART TWO

My Collaboration with Herbert Dingle

As mentioned earlier, I exchanged a few letters with Herbert Dingle when I was in Cambridge in 1971-72, but I did not engage in any very serious discussion with him. I acquired a copy of *Science at the Crossroads* in January 1973, and after reading it I wrote to him again. This renewed correspondence continued until just before his death in 1978, and ranged over many subjects. In *The Relativity Question* (McCausland, 1988) I described many features of our collaboration, and included verbatim copies of a good deal of correspondence between him and people whose help he sought in attempting to present his thesis more effectively.

Chapter 4 is a very brief biography of Herbert Dingle, with particular reference to the development of his views on relativity. Chapter 5 is a brief summary of his book *Science at the Crossroads*, and Chapter 6 is the story of how two very influential scientists reacted to his campaign.

Chapter 4

Herbert Dingle

> Dingle is one of the very few honest (former) relativists. He was a fervent admirer of Einstein and used to look at me with kindly contempt for not being able to understand the greater mysteries. At last I got him to 'see the light', and as you know, he is now the most persistent opponent.
>
> G. Burniston Brown, letter to Ian McCausland, February 1974.

Herbert Dingle (1890-1978) was a very eminent British scientist. The best biography of him that I am aware of is an obituary article by a colleague who knew him well (Whitrow, 1980). I shall not attempt to give a comprehensive account of his life and accomplishments, but merely give a brief summary of his life, especially as it related to relativity.

Herbert Dingle was born in London on 2 August 1890. After the death of his father when he was quite young, his mother moved to Devonshire, and Dingle received his early education at the Plymouth Science, Art and Technical Schools, which he left when he was fourteen. He worked as a clerk for 11 years, continuing to study in his spare time, and eventually won a Royal Scholarship for Physics at the Imperial College of Science and Technology in London. He received his B.Sc. degree from the Imperial College of Science and Technology, London, in 1918, and the D.Sc. degree in 1930. After his graduation in 1918 he was successively Demonstrator, Lecturer, Reader and Professor of Natural Philosophy at Imperial College, during the period 1918-1946; he then became Professor of the History and Philosophy of Science at University College, London, a position which he occupied until becoming Professor Emeritus in 1955. He died in Hull on 4 September 1978.

It was in the year following Dingle's graduation from Imperial College that Eddington's famous eclipse expedition, which was responsible for the rapid rise to world-wide fame of Albert Einstein and his theory of relativity, took place. It was therefore natural for Dingle to develop an interest in relativity. He described one of his early articles on the subject, and his first impressions of Einstein, in the book *Einstein: the man and his achievement* (Whitrow, 1967). He wrote that he had first seen Einstein in June 1921 when he came to England and gave a lecture on relativity at King's College, London. Dingle had written an article on relativity for the *Westminster Gazette*, then one of several London evening newspapers, which appeared on the day of Einstein's visit, and in recognition of this he had received a Press ticket for the lecture. He described how the occasion allowed him to form a more balanced view of the significance of Einstein and his theory; this is how he described the lecture:

> It was delivered in German and my consequent inability to follow it allowed me to take undisturbed advantage of my favourable situation to observe the phenomenon who had, as it then appeared, displaced Newton as the unique interpreter of the fundamental laws of the universe. No one could live up to the expectations that the atmosphere of the time had engendered. Instead one saw a human being, to all appearances modest and serene—not the unfolder of secrets hidden from the beginning of the world but the initiator of a natural development in the progress of human understanding. It was a salutary corrective.

Dingle's interest in relativity continued to develop, and in 1922 he published a book on relativity for the general reader (Dingle, 1922). As Whitrow said in his obituary, this book showed both his grasp of the subject and his powers of lucid exposition. In November 1928 he published a leading article in *Nature* (Dingle, 1928) on the occasion of the ninth anniversary of the historic meeting at Burlington House at which the eclipse results were presented. That article aroused considerable interest and led to discussion that continued to be published until February 1929.

In 1932 he spent a year as Rockefeller Foundation Scholar at the California Institute of Technology, where he had the opportunity to meet Einstein in person.

It is interesting to study the evolution of Dingle's views on the behaviour of clocks in both special and general relativity. An early example of his view is described as follows in his book *Science and Human Experience* (Dingle 1931, 68):

> Relativity teaches us that if we were to travel about the universe at high speed we would discover, on returning to the Earth, that a longer time had elapsed than our perfectly accurate clocks would indicate.

Then he published (Dingle, 1939) a criticism of a paper in which Dr. Herbert E. Ives had described what he called "the physical and logical absurdity of the 'clock paradox'." The discussion centred around the case of two similar clocks, in which one moved to a distant point at high velocity and returned to the other. Here is part of what Dingle wrote:

> If, then, the clocks show different times at the end of the experiment, they must have been treated differently; and obviously they have. Clocks do not naturally move apart and come together again: in order to make them do so, some force must be applied to one and not to the other, and the difference in reading must therefore be due to the difference in physical treatment. Relativity does not predict what clocks will do when constrained to move unnaturally: it simply says that the *motions* involved may be described with equal validity with respect to any assumed standard of rest. [Italics in the original.]

Although Dingle showed that he believed that it was possible for asymmetrical ageing to occur in the case that he was discussing, I think it is noteworthy that he does not appear to have believed in 1939 that asymmetrical ageing followed from the special theory, but rather that the theory had nothing to say about the matter.

An interesting outcome of that paper of Dingle's was an encounter with an earlier opponent of relativity, Professor Alfred O'Rahilly (1884-1969), of University College, Cork, who had published a book (O'Rahilly, 1938) in which he was highly critical of relativity. He wrote to Dingle in June 1939 to ask him whether he would be willing to submit a paper to *The Philosophical Magazine* for him. Although it was obvious from O'Rahilly's letter that the proposed article would disagree with what Dingle had written in the paper mentioned above, he readily agreed to submit the article for O'Rahilly. In his letter of 30 June 1939, transmitting the article to Dingle, O'Rahilly expressed his thanks by saying "I very much appreciate your courtesy to a complete stranger. You see, I am an obscure little heretic, while you are big enough at times to appear as a heresiarch." He also included the following paragraph:

> Excuse my impertinence in adding a personal remark. I have often been stimulated by your sceptical attitude & by your fearless criticisms of bigwigs. But I find it awfully hard to understand how you swallow so much of the stuff that is called 'relativity'. I feel that it is in this region that the courageous bluntness of a man of your standing is most needed. Please don't be offended.

Unfortunately O'Rahilly's paper was not published in *The Philosophical Magazine*. It is noteworthy, however, that Dingle was willing to transmit O'Rahilly's paper to the journal, even though he knew in advance that it would be critical of his own paper, but when Dingle later tried to publish papers critical of relativity, he received very little support from other scientists for his attempts.

Later, when other scientists claimed that asymmetrical ageing followed from the special theory, Dingle did not accept their claim. He was prompted to publish a criticism of that interpretation of the theory by the appearance of the book *The Foreseeable Future* (Thomson, 1955) that suggested that "the best opinion is that the time contraction would occur." His criticism (Dingle, 1956) appeared in *Nature* with the title "Relativity and Space Travel." Although he still believed that the special theory was sound, he wrote that Einstein's original 1905 paper on the subject "contains a most regrettable error, in a statement that a clock moving in a closed curve will be found, on returning to its starting point, to be behind a stationary clock."

Since Dingle's article (Dingle, 1956) criticized an earlier argument of his (McCrea, 1951), McCrea was given the opportunity to publish a reply in the same issue of *Nature* as Dingle's article, and his reply was followed by a further reply by Dingle, also in the same issue; so started the public debate between Dingle and McCrea on the clock or twin paradox. Professor McCrea (1904-1999) was a Professor at Royal Holloway College, University of London, and had, as mentioned earlier, written a book on Relativity (McCrea, 1935). He and Dingle knew each other well, and carried on the debate by private correspondence as well as in their published papers. I have read some of the correspondence in Dingle's papers in the archives of Imperial College, and can attest to the friendliness of their correspondence. One amusing item is related to the fact that McCrea's

1956 reply to Dingle was sent from the University of California at Berkeley, where he was then visiting: in a letter to McCrea, dated July 6 of that year, Dingle wrote: "I hope you are now safely back from the land of the Yanks, and have left them wiser (take this both ways)."

The above-mentioned published items led to a vigorous discussion both in the scientific literature and in more popular writings. Since there has been such an enormous amount of published discussion about the clock paradox and the twin paradox, I shall not attempt to discuss them further here. As mentioned earlier, a very comprehensive survey of the discussion can be found in the book *Time and the Space Traveller* (Marder, 1971). Also, a long and comprehensive article on Dingle and the clock paradox was written by Hasok Chang (Chang, 1993).

I think it should be pointed out that, at the time that Dingle was criticizing the asymmetrical ageing predicted by the special theory, he remained an admirer of the theory and of Einstein, as described, for example, in the excerpt from a letter from Dr. G. Burniston Brown that I have used as an epigraph to this chapter. Dr. Burniston Brown was a long-time critic of relativity, and wrote an important paper criticizing it, which is discussed in Chapter 14.

Herbert Dingle's scepticism about the clock paradox eventually led him to the conclusion that the special theory contained a fatal contradiction. Clearly, if the special theory is wrong, the clock paradox, which arose from the theory, becomes much less important. It is unfortunate that, because of the prominence of the clock-paradox controversy in the late fifties, it is this controversy that is still linked with Dingle in many people's minds. Many writers continued to criticize his arguments as if he was still arguing against the orthodox resolution of the paradox, despite explicit statements to the contrary in his book *Science at the Crossroads*. In fact, it was his attempt to convince the scientific world that the special theory was wrong that occupied much of his time and energy during the last twenty years of his life, and it is that problem with which I am concerned in the present book. Although I am not greatly concerned with the clock paradox, there is one strong similarity between that controversy and the controversy over the validity of the theory, namely the diversity of the replies that have been made in defending the orthodox point of view. This diversity, in the case of the clock paradox controversy, was aptly described by a Professor of Electrical Engineering (Cullwick, 1959) in the following words:

> On one thing Professor Dingle's critics are all agreed, that he is wrong. They do not all agree, however, on the nature of his error. Some give arguments which are no more than illustrations of the obvious fact that the reciprocal Lorentz transformation is algebraically consistent; some claim that the problem requires the General Theory of Relativity; and some appear to regard the matter as settled by their knowledge of four-dimensional space-time. Some argue with patience, while others thinly disguise their irritation.

After mentioning some of the diverse opinions on the subject, Cullwick continued as follows:

> One is reminded a little of the battle of Arsuf, in the Third Crusade, when, led by Richard, the crusaders routed the infidel with much blood and satisfaction and then started to slay each other.

While the comparison with the battle might have been appropriate in the case of the controversy on the clock paradox, it is not a good comparison for the controversy on the validity of the special theory. In the latter controversy the different defenders of the theory are, indeed, inconsistent with one another in their arguments, as in the former case. They do not, however, attack the arguments of other defenders; they even support them, presumably on the basis that "my enemy's enemy is my friend." For example, as I describe in Chapter 6, G.F.R. Ellis complimented Ziman's review of *Science at the Crossroads* as being "admirable" but went on to contradict his argument.

The start of Dingle's claim that the special theory is inconsistent is more difficult to identify precisely than the start of his criticisms of the clock paradox, partly because his early attempts to show that the theory was inconsistent were rejected by editors. He himself (Dingle, 1962) identified one paper (Dingle, 1960) as his first published claim to an inconsistency. Then in 1961, in the book *A Threefold Cord* (Samuel and Dingle, 1961), he described some of the difficulties that he had encountered in attempting to have his criticism published by The Royal Society, the Physical Society, *The Philosophical Magazine*, and *Nature*. For example, in one of the more striking examples of the attitude of a scientific journal, Dingle described how *The Philosophical Magazine* sent back a critical paper by return mail, with a statement that subjects of a polemical nature were not suited to that journal! After describing that rejection, Dingle went on to say; "One of the leading scientific journals will not publish anything 'of a polemical nature', which can only mean that, in science itself, it will not publish any criticism of orthodox views. Accept them, and your paper will be considered for publication; question them, and it will not."

In Appendix II of *A Threefold Cord*, Dingle presented a "Proof that Einstein's Special Theory cannot correspond with fact," which contains a very strong argument in favour of his claim of an inconsistency in the theory. As he later pointed out (Dingle, 1972, 41), no reviewer of *A Threefold Cord* even mentioned that the question of the validity of special relativity had been raised. Since I believe that the argument that he published there is sound, and to the best of my knowledge has not been refuted, I include it with some added commentary as Appendix I of this book.

In the same year as the publication of *A Threefold Cord*, Dingle published a second edition of his short textbook *The Special Theory of Relativity* (Dingle, 1961a), which he had originally published in 1940. When he revised the book for the second edition, he had already become convinced that the theory was untenable, and he included in the book a statement of that point of view. He pointed out in the Preface to the 1961 Edition that since the book had first appeared he had become convinced, for reasons that he believed conclusive, that the theory was no longer tenable. He explained that, although his first impulse was to withdraw the book from circulation, on second thoughts he decided that it would be more fitting to re-issue it with an explanation, which he then proceeded to give.

Dingle's decision to re-issue the book, even though he believed that the theory was no longer tenable, had some embarrassing consequences for at least one physics department. There is an interesting description of the "adoption" of the book at Harvey Mudd College, in Claremont, California, on a website* devoted to "Earliest History of the HMC Physics Department." It appears that in those days few institutions included relativity in their introductory physics course, and the instructor who planned to teach the course had to find a suitable textbook. He found the earlier edition of Dingle's book in the library, and it seemed to be just right for the course, so enough copies of the book were ordered for the class. When the books arrived, the instructor discovered that they were the later edition, with Dingle's statement that he believed that the theory was

* http://www.hmc.edu/academicsclinicresearch/academicdepartments/physics/history/earliest.html

no longer tenable, not the edition he had seen in the library. By that time it was too late to find another book, so the instructor had to write some notes of his own, which eventually grew into a new textbook (Helliwell, 1966).

There is a great diversity among the replies that have been made to Dingle's claim that there is a contradiction in the special theory. Despite the fact that some of Dingle's critics contradict each other, some contradict Einstein, and some even contradict themselves, few scientists seem to be concerned about the contradictions, and Dingle's critics still seem to be unanimous on only one thing—that Dingle is wrong. The first time that I became aware of the diversity was in the correspondence in *The Listener* that went on in 1971-72, that I mentioned in Chapter 1. I described some of the contradictions in my article "Problems in Special Relativity" (McCausland, 1983b), which is reproduced as Appendix II of the present book.

Before leaving this brief biographical sketch of Herbert Dingle, I should say something about his relationship with the journal *Nature*. During the editorship of Sir Richard Gregory, from 1919 to 1938, Dingle was very active in publishing in the journal, and in 1928 he was entrusted by Gregory to write the editorial (Dingle, 1928) on "The Understanding of Relativity" which was mentioned above. He was also part author of the biography of Sir Norman Lockyer, the first editor of *Nature* (Lockyer, 1928).

At the time of the Silver Jubilee of King George V in 1935, Dingle was the author of one of the papers in a special issue of *Nature*, dated May 4, 1935, along with such eminent scientists as Sir James Jeans, Sir William Bragg, Lord Rutherford, Professor F. A. Lindemann (later Lord Cherwell) and Professor E. V. Appleton. The Silver Jubilee is one of my earliest memories; I was taken by my parents to a bonfire celebrating the event, shortly after my sixth birthday.

When Sir Richard Gregory retired, Dingle wrote a very touching letter to him which is published in Armytage (1957). He expressed his gratitude to Gregory for his cooperation and support, and referred to "the sense of mingled power and security which has come from the consciousness that *Nature* has provided, at the same time, the freest medium of expression and the highest wisdom to check indiscretions which the limitations of human nature allow."

On Gregory's retirement as editor, Dingle declined an opportunity to become the next editor, as he described in a January 1974 letter to the then editor, David Davies: "When, on Gregory's retirement, I was sounded as to whether I would consider succeeding him if formally invited, I declined with the speed of light."

During the period of my collaboration with Herbert Dingle, the two editors of *Nature* were John Maddox, who was editor from 1966 to 1973, and David Davies, editor from 1973 to 1980. John Maddox was editor again from 1980 to 1995, after Dingle's death.

After several years of attempting to obtain an answer to his criticism of special relativity, Dingle became so convinced of the moral shortcomings of the scientific community, in its reluctance to meet or to answer his criticisms of the special theory, that he eventually published *Science at the Crossroads* in an attempt to draw the attention of the scientific community and the general public to what he considered to be a highly unsatisfactory state of affairs. That book is discussed in Chapter 5.

Herbert Dingle kept up his campaign against special relativity until just before his death. As I mentioned before, I corresponded with him for several years before his death, and I also corresponded with Mr Mark Haymon, a lawyer who supported him in his quest for an answer to his question about special relativity. From them I received copies of many letters that they had written and received in the course of their activities, and it was understood that I would try to publish that information if they were unable to do so themselves. As I describe later in this book, I did publish some of those letters in *The Relativity Question* (McCausland, 1988) and, as I describe in Chapter 7, I also arranged for the publication of Dingle's last scientific paper (Dingle, 1980) in *Wireless World*. Another posthumous paper was "Time in Philosophy and in Physics" (Dingle, 1979).

In addition to corresponding with Dingle for some years, I was able to meet him in person on two occasions. The first was in July 1974, in London; the second in August 1978, in Hull. Unfortunately, on the second occasion, he had a stroke the evening before I arrived, so my visit was to the hospital to which he had been taken. Although it was very sad to find him in that condition, I was glad to see that his mind and his sense of humour were still active, and that he was still interested in the relativity problem. He died on 14 September 1978, just four weeks later.

After Herbert Dingle's death, I looked forward to reading what journals like *Nature* would write about him. I found it interesting that the first mention of him that I found in *Nature* after his death was in a review in the 9 November issue (Rowan-Robinson, 1978), in which the reviewer compared Dingle unfavourably with Immanuel Velikovsky (1895-1979). Having written that Velikovsky behaved with considerable dignity and that that was not always true of those who were rejected and despised by the scientific establishment, he continued: "In some, like Herbert Dingle in *Science at the Crossroads*, the bitterness of their treatment becomes more important than the scientific case they are trying to make." Since Rowan-Robinson was on study leave in California at the time his review was published, he may not have known of Dingle's death when he wrote the review, but if he did not know of it he presumably intended Dingle to see his gratuitous insult. The editor, David Davies, certainly knew of Dingle's death when he published the review. Rowan-Robinson would not have known in 1978 that Velikovsky had also written a forceful story of his own struggles, with copious documentation of the antagonism of the scientific world towards his writings; this was published posthumously (Velikovsky, 1983).

Nature then published an obituary of Herbert Dingle in the 15 February 1979 issue, (Whitrow, 1979). It was noteworthy that it contained a rebuttal of Dingle's claim that special relativity was self-contradictory. I found it rather strange that the editor, David Davies, allowed such a rebuttal to be published, especially as he had told me in July 1977 that "I cannot think of anything that needs to be said which hasn't already been said." (I have described my correspondence with Davies in *The Relativity Question* (McCausland, 1988, 28-9), so I don't think it is necessary to describe it again.)

In his other obituary of Dingle, which I mentioned at the beginning of this chapter, Whitrow (1980) again published a rebuttal of Dingle's argument. In both these obituaries he wrote about what he called Dingle's "implicit requirement" and ended his argument with the following sentence, identical in both publications: "Dingle's requirement is therefore equivalent to adopting the Newtonian concept of universal time, and this is incompatible with special relativity."

This last sentence is a clear example of begging the question, since the writer is simply saying that Dingle's argument is wrong *because it is incompatible with special relativity*. Since it is the validity of special relativity that is at issue, incompatibility with the theory should not be used as a criterion to claim that Dingle's argument is unsound. I wrote to *Nature* to point this out, but of course my submission was refused publication.

There was an interesting anonymous obituary in *The Times* (London) on September 6, 1978. Then in the September 18 issue there appeared a supplementary note by Professor Whitrow. One sentence is interesting; referring to Dingle's campaign against special relativity, he wrote: "Although I, for one, did not see eye to eye with him, I had no sympathy whatever for the scurrilous abuse to which he was at times subjected by some scientists and engineers not professionally expert in the subject." His objections to the scurrilous abuse were apparently not strong enough to make him refrain from publishing two obituaries in which he rebutted Dingle's argument, when he was unable to answer back.

Chapter 5

Science at the Crossroads

> This most readable book, written with characteristic lucidity in beautiful English, conveys all the liveliness of mind and personal charm recognized by everyone who has had the privilege of meeting Professor Dingle.
>
> J. L. White (1973).

> Suppose I were to try and prove that the earth is flat and then try to get this accepted by the scientific community; the catalogue of rebuffs would provide the basis for a book like Dingle's.
>
> Roy Johnston (1973).

Herbert Dingle's book *Science at the Crossroads* (1972) described in great detail the history of the controversy up to that time, and the ways in which some members of the scientific community had responded to his criticisms of special relativity. Although I believe that it is necessary to read the book if one is to acquire a thorough understanding of the controversy, the following very brief sketch of the book is included here in order to make the remainder of the present narrative intelligible to those who have not yet read it.

One of the things that Professor Dingle emphasized very strongly in the book was the fact that it was the validity of the special theory of relativity that was at stake, not the much less important problem of the resolution of the clock paradox or twin paradox. In his Introduction he wrote:

> I can present the matter most briefly by saying that a proof that Einstein's special theory of relativity is false has been advanced; and ignored, evaded, suppressed and, indeed, treated in every possible way except that of answering it, by the whole scientific world (the world of *physical* science, that is; the theory has no place at present in the biological and psychological sciences).

The main body of the text of *Science at the Crossroads* is divided into two parts, called *The Moral Issue* and *The Intellectual Issue*. In *The Moral Issue* Dingle presented a factual narrative, documented by many quotations from his interlocutors, describing the responses of various named members of the scientific community to his attempts to obtain an answer to his Question, a crucial question on which hinged the validity of the special theory. The wording of the Question appears later in this chapter.

Dingle described how he made a second attempt to have his criticism published by the Royal Society (the first attempt having already been described in Samuel and Dingle (1961), as mentioned in Chapter 4). This new paper was rejected on the recommendation of two referees. Although one of the referees stated that the paper contained an elementary fallacy, Dingle was unable to obtain from the Royal Society a statement of what the alleged fallacy was. The other referee wrote that "the Society would make itself ridiculous by publishing this paper."

It is interesting to compare the reports of the two Royal Society referees with the letter that Professor P.M.S. Blackett (1897-1974) wrote to Dingle, which appears on pages 96-7 of the book. The letter was written in reply to Dingle's request that Blackett read his paper "and, if he thought it worthy of consideration, submit it to the Royal Society for publication." In the letter, Blackett, who won a Nobel prize for physics in 1948, became President of the Royal Society in 1965, and was created a life peer in 1969, wrote that he was "naturally interested in relativity, particularly as I have taught the special theory for many years to our first-year students." He went on to say "I confess I cannot completely follow the details of yours and Max Born's argument, or rather, I have not had the time or inclination to do so. But to my superficial knowledge of the subject, there is nothing obviously wrong with the ordinary formulation."

It appears that the eminent scientists of the Royal Society who rejected Dingle's criticisms could not make up their minds whether his criticisms were too trivial to be taken seriously, or too difficult for a long-time teacher of the theory to understand.

Dingle's experience with the Royal Society might be compared with that of another author, John Waterston, who in 1845 submitted to the Royal Society a paper on the kinetic theory of gases, with the title "On the Physics of Media that are Composed of Free and Perfectly Elastic Molecules in a State of Motion." As described by Haldane (1928), the paper was rejected by two referees, one of whom wrote that "the paper is nothing but nonsense, unfit even for reading before the Society." The manuscript was discovered nearly fifty years later in the Archives of the Royal Society by Lord Rayleigh, who had it published in the *Philosophical Transactions*. Rayleigh wrote an introduction to the paper, in which he stated that the omission to publish the paper when it was submitted was a misfortune which probably retarded the development of the subject by ten or fifteen years.

According to the online encyclopaedia Wikipedia, in the entry for John James Waterston, the referee who wrote that Waterston's paper was "nothing but nonsense" was Sir John William Lubbock (1803-1865), who was Vice-President of the Royal Society at the time. It is obvious that high officials of the Royal Society are capable of making mistakes, and perhaps the referees of Dingle's paper to the Royal Society might also have been mistaken. If his paper is still in the Archives, it would be possible for the Society to look again at the paper, and to provide a good reason for its rejection. As in the case of Waterston, fifty years late would be better than nothing.

In view of later developments that I describe in the next chapter, I believe it is important to mention some correspondence between Herbert Dingle and Mr. John Maddox, the Editor of *Nature*, that is discussed in the book. There had been a long debate on relativity in *The Listener* in 1969, resulting from an article (Dingle, 1969) in the July 3 issue, and after the discussion ended Dingle had sent a copy of the debate to Maddox. As a result, Maddox wrote to Dingle on 24 November 1969, stating that he proposed to write a leading article summarizing the position, and that he would publish it "before the end of the year." It did not appear before the end of that year, and in response to an enquiry the Editor wrote to Dingle on 21 January 1970 to say that the article was "almost ready." Toward the end of March another enquirer, Lord Soper, wrote to Mr. Maddox, and was told that it would be "a week or two" before the article was ready; when Lord Soper enquired again on 6 July, he received no reply. The promised leading article was never published, and I discuss in Chapter 6 the reasons subsequently given by Maddox for its non-appearance.

Professor Dingle's book also contains an interesting historical survey of the development of relativity theory, and the relationship between Einstein's special theory and Lorentz's theory, the latter theory being quite different from Einstein's in that it assumes a stationary ether, such that clocks moving through the ether would *actually* run slower than clocks that remained at rest. Dingle pointed out that the two theories are often confused with one another, and stated that all the experimental evidence that is taken to support Einstein's special theory could, with equal validity, be taken to support Lorentz's quite different theory. I discuss this subject in Chapter 10.

Dingle also pointed out, both in his book and elsewhere, that the experimental evidence that is taken to support the special theory depends on circular arguments, since it relies on the validity of Maxwell's electromagnetic theory to infer certain intermediate results such as the velocities of certain elementary particles. I discuss these points in more detail in Chapter 11.

Professor Dingle's conviction that there is a fatal flaw in special relativity arose from the scepticism that had been aroused by the clock paradox. As he pointed out in his book, a paradox arises when, from the same premises P, two apparently contradictory conclusions, X and Y, seem inescapably to follow. Such a paradox can be resolved if and only if one of the following things can be shown:

(1) the conclusions are not really contradictory,
(2) conclusion X does not follow,
(3) conclusion Y does not follow,

or

(4) the premises P contain a contradiction.

Suppose, for example, that there are two identical clocks A and B, initially together and mutually synchronized. Suppose that A moves away from B at uniform speed, and later turns around and returns to B at the same speed. In terms of the notation above, assume that the premises P are the axioms and definitions on which the special theory of relativity is based, conclusion X (symmetrical ageing) is that the readings of A and B are equal at the reunion of the two clocks, and conclusion Y (asymmetrical ageing) is that the readings of A and B are unequal at their reunion. Clearly X and Y are contradictory, ruling out possibility (1) of the first paragraph. As Dingle pointed out, Einstein in his original paper accepted conclusion Y but did not disprove conclusion X. It should be emphasized that all additional proofs of either X or Y do nothing to resolve the paradox, because any such proof does not disprove the other result; since Dingle was unable to disprove Y to his complete satisfaction, he was eventually forced to consider the possibility that the paradox could only be resolved by finding a contradiction inherent in P. Once he had found what he believed to be a contradiction in the theory, he tried to find ways of expressing the contradiction in such a way as to avoid the accelerations which are inevitable in any experiment in which two clocks, or twins, separate and later reunite.

By the time of the publication of *Science at the Crossroads* in 1972, Dingle had refined his thesis in such a way that it could be expressed in two ways, The Argument and The Question. The Argument is presented on page 45 of *Science at the Crossroads*, in the following words:

THE ARGUMENT

> According to the special theory of relativity, two similar clocks, A and B, which are in uniform relative motion and in which no other differences exist of which the theory takes any account, work at different rates. The situation is therefore entirely symmetrical, from which it follows that if A works faster than B, B must work faster than A. Since this is impossible, the theory must be false.

The Question might be worded very briefly as follows: *Which of two clocks in uniform relative motion does the special theory require to work more slowly?* However, in order to present the story satisfactorily we should consider The Question in its extended form, as it is presented on pages 45-46 of *Science at the Crossroads*:

THE QUESTION

> According to the special relativity theory, as expounded by Einstein in his original paper, two similar, regularly running clocks, A and B, in uniform relative motion, must work at different rates. In mathematical terms, the intervals, dt and dt', which they record between the same two events are related by the Lorentz transformation, according to which $dt \neq dt'$. Hence one clock must work steadily at a slower rate than the other. The theory, however, provides no indication of which clock that is, and the question inevitably arises:

> How is the slower-working clock distinguished? The supposition that the theory merely requires each clock to *appear* to work more slowly from the point of view of the other is ruled out not only by its many applications and by the fact that the theory would then be useless in practice, but also by Einstein's own examples, of which it is sufficient to cite the one best known and most often claimed to have been indirectly established by experiment, viz. 'Thence' [*i.e.*, from the theory he had just expounded, which takes no account of possible effects of acceleration, gravitation, or any difference at all between the clocks except their state of uniform motion] 'we conclude that a balance-clock at the equator must go more slowly, by a very small amount, than a precisely similar clock situated at one of the poles under otherwise identical conditions.' Applied to this example, the question is: what entitled Einstein to conclude *from his theory* that the equatorial, and not the polar, clock worked more slowly?

After his statement of the above question, Dingle went on to say that the remainder of the book was concerned with The Question and the attitude of the scientific world to it. In his Introduction (p. 18) he wrote that any reviewer of the book could dispose of its basic *raison d'être* by giving an answer to the question.

One of the subjects that Dingle discussed was the relation between Einstein's theory and Lorentz's similar theory, and the fact that the two theories had similar status before the rise of Einstein's General Theory raised the fortunes of the Special Theory too. I should say that he unfortunately exaggerated some of the evidence for the historical relation between the two theories, when he compared two books by Ebenezer Cunningham. Dingle pointed out that, in the first book (Cunningham, 1914), Lorentz had 13 entries in the index and was mentioned throughout the book, but in the second book (Cunningham, 1921), the name of Lorentz did not appear at all in the index. Dingle interpreted this as a transfer of credit from Lorentz's theory to Einstein's. However, unfortunately for this argument, Einstein's name does not appear in the index of the second book either! In the first book Einstein had 10 index entries to Lorentz's 13; in the second neither name appears in the index, just because the indexing was done differently in the two books.

One of the prominent features of *Science at the Crossroads* is Dingle's repeatedly expressed fear that some terrible accident might occur because of unjustified faith in a flawed theory. I do not wish to dwell on this feature of his argument. I believe that the importance of scientific truth, or at least the search for truth, is enough to justify the scientific community's obligation to answer Dingle's criticisms of special relativity.

In Chapter 1 of the book Dingle discussed the two possible ways in which a theory might be overthrown: by experience, or by reason. He described the big difference between the two on page 25 by saying "Experience—*i.e.*, observation or experiment—usually carries much greater conviction than reason, though ultimately they have equal authority." He pointed out that "there has frequently been less readiness to abandon a cherished idea on rational than on experimental or observational grounds." This is indeed true in the history of this controversy, since some defenders of relativity have claimed that it is necessary to find a new theory to replace the theory that is being criticized, even if the criticism is fatal.

Although there were other active critics of relativity, such as Dr. G. Burniston Brown (1902-1988) and Dr. L. Essen (1908-1997), who were active while Herbert Dingle was writing his book, he did not discuss their ideas since he thought it best to concentrate on the one central question of his book. In the same way, I think it best to concentrate on one issue and not to discuss the ideas of other contemporary critics.

Dingle considered the possibility of using Ritz's electrodynamic theory instead of Maxwell's theory, which is now the established theory of electromagnetism. Unfortunately, in his description of that theory, Dingle made some rather confusing statements, and I would like to try to clear up the problem that these statements may have caused. I should emphasize that I am not advocating Ritz's theory as a replacement for Einstein's; I am simply trying to clear up some confusion.

According to the second postulate of special relativity the velocity of light is independent of the velocity of its source, but in Ritz's theory light has velocity c with respect to its source. That means that, if a source that is moving at velocity v relative to some reference frame emits a pulse of light, that pulse would have velocity $c + v$ relative to that reference frame.

Unfortunately there is an ambiguity if the source, after emitting a light pulse while travelling at velocity v relative to the reference frame, changes its velocity from v to w. There are two possibilities, which I shall distinguish as follows:

(a) the velocity of the light pulse relative to the reference frame remains at $c + v$

(b) the velocity relative to the reference frame changes from $c + v$ to $c + w$

The latter possibility, that the velocity of light might change if the velocity of the source changed after the light had left the source, may seem rather far-fetched, but I believe it is compatible with an idea of Faraday's, that each source has a system of rays emanating from it and that the vibrations on these rays constitute light. A simple two-dimensional model of this idea is a source of light represented as the hub of a wheel, the spokes of the wheel representing the rays, the light considered as a wave motion along the spokes, and the wheel moving about as a whole when the hub moves. I published a paper discussing this idea (McCausland, 1992), which is reproduced in Appendix III.

Dingle discussed the above possibilities (a) and (b) in a correspondence with professor E.G. Cullwick in 1958, and he appeared to be in two minds about which of the two possibilities Ritz's theory entailed. When he considered possibility (a), that the velocity of the light pulse keeps the velocity that it had at emission, he wrote, in a letter to Cullwick dated 1958 October 21, "But with respect to what? The emitting body may be the only thing in the universe."

The confusion that I think needs to be cleared up occurs on page 205-7 of *Science at the Crossroads*. There Dingle discussed an argument, originally put forth by de Sitter, that the light from double stars (which rotate around their common centre of gravity) would show a totally confused picture of the stars if the light travelled at velocity c relative to the source. Dingle discussed the light from a hypothetical double-star system in which two stars rotate about their common centre of gravity, and he assumed that the stars were one light-minute apart and 100 light-years from the earth. According to special relativity, the light beams from the two stars would reach the earth not more than one minute apart. Here is what Dingle wrote on page 207:

> If we consider the same observations on the supposition that each beam of light moves throughout with velocity c with respect to its own source, it is at once evident that after any time (duration) t, however great, each beam will be at distance ct from its own source, and therefore the beams can never be further apart than their sources, *i.e.*, 1 light-minute.

This statement is confusing because Dingle did not state which of the two forms (a) and (b) he was using. I think that most readers would assume that he meant alternative (a) of the two possible interpretations given above, and according to that alternative the above-quoted statement is wrong. What I believe he should have said is that, after any time (duration) *t*, each beam would be at distance *ct* from where its source *would have been* if it had kept on moving at the same velocity as it had when the light was emitted from it. However, according to alternative (b), his statement is correct.

I hope that this clears up the confusion that Dingle introduced by failing to clarify which interpretation of Ritz's theory he had in mind when he wrote his book. I think I should also point out that there are other possible explanations of the observations of double stars that do not require Einstein's second postulate to be true, one of which I described in a short note (McCausland, 1980).

I think it is interesting to compare *Science at the Crossroads* with Lee Smolin's book *The Trouble with Physics* (Smolin, 2006) which might have been a better title for Dingle's book than the one he used, because the trouble he described was in physics rather than science in general. However, both books describe a dogmatic and emotional adherence to a theory, special relativity in Dingle's case, and string theory in Smolin's. Smolin (2006, 310) described the difference between two groups of scientists, the master craftspeople and the seers. The former go into science "because, for the most part, they have discovered in school that they're good at it." The seers are dreamers who go into science "because they have questions about the nature of existence that their schoolbooks don't answer." I believe that Herbert Dingle was a seer, and that most of his critics were craftspeople who were unwilling to accept that mathematical consistency of the Lorentz transformation is not a sufficient condition for logical consistency of the special theory.

Several reviews and articles about *Science at the Crossroads* appeared soon after it was published. I described some of the features of several of the reviews in my article that appears in Appendix II (McCausland, 1983b). In the next chapter I discuss in more detail the way in which two eminent scientists behaved toward the book and toward Dingle himself.

Chapter 6

Einstein's Champions

> The objective of science is not just to acquire information nor to utter all non-contradictory notions; its goal is a consensus of rational opinion over the widest possible field.
>
> John Ziman, (1968).

> "... consensus-building is inherently a political process whose practitioners are called 'spin-doctors' in other trades."
>
> John Maddox, (1996).

Two scientists in particular were very active in the story of Herbert Dingle's campaign against relativity, John Maddox and John Ziman. In my opinion, they were the two who were most active in preventing Dingle's criticisms of special relativity from being given enough exposure to be answered rationally.

John Maddox (1925-2009) was the Editor of *Nature* from 1966 to 1973, and again from 1980 to 1995. In view of the fact that one of the most striking passages of *Science at the Crossroads* is Herbert Dingle's account of Maddox's failure to publish a promised leading article, it is interesting that one of the earliest published comments on the book was a leading article in *Nature* (Anon., 1972). Although the article was anonymous, it was fairly obvious that John Maddox was the writer.

The leading article about Dingle's book is such an incoherent mess that it is difficult to know where to begin in assessing it. Let us begin by quoting the first sentence and the last two sentences of the article, which are:

> Everybody is fond of Professor Herbert Dingle, as well as of the clock paradox in special relativity which he has single-handedly nurtured since the early 1930s.
>
> And is there any hope that he will now be satisfied with the demonstration that moving clocks run at different speeds from clocks at rest which has been provided in the past few months by the experiment in which Hafele and Keating have flown caesium clocks in different directions around the world (*Science*, **177**, 166; 1972, see also *Nature*, **238**, 244; 1972)? It will be sad to see the clock paradox disappear, but this work is the last nail in the coffin.

The writer did not appear to have noticed Dingle's statements in the book that he had for years held an open mind on the subject of asymmetrical ageing, or his attempt to make clear that the scientific issue was not what is normally associated with the expression "clock paradox"; as mentioned in Chapter 1, these statements are found in the Preface of *Science at the Crossroads.* Furthermore, the expressions "single-handedly" and "since the early 1930s" in the first sentence of the article are both totally inaccurate. The appeal to the result of the Hafele-Keating experiment is discussed in Chapter 11, where I deal with the question of experimental evidence for the special theory.

The article quoted part of The Question (Dingle, 1972, 45-46) with added comments, in the following form:

> "According to the special relativity theory ... two similar, regularly running clocks ... in uniform relative motion must work at different rates. Hence one clock must work steadily at a slower rate than the other. The theory, however, provides no indication of which clock that is, and the question inevitably arises: How is the slower-working clock distinguished? The supposition that the theory merely requires each clock to appear to work more slowly from the point of view of the other is ruled out merely by its many applications and by the fact that the theory would then be useless in practice, but also by Einstein's own examples ..." The trouble, of course, is that in the last of these sentences, Dingle is denying the central principle of relativity. And why should he not accept that each of two clocks in uniform relative motion should appear to run slow from the other's point of view? That, according to the relativists, is what the real world is like. [All the ellipses are as they appear in the

quotation; there is a small error in that the second appearance of the word "merely" should read "not only."]

If Dingle was "denying the central principle of relativity," as the article suggests, he did it by referring (in the part of The Question that Maddox replaced by the ellipsis after "Einstein's own examples") to Einstein's prediction from the special theory that a clock at the equator would work (not just *seem* to work) more slowly than a clock at one of the poles. (See my quotation of The Question in Chapter 5, where the full sentence can be found.) Now, if Einstein was correct in his deduction from the theory that an equatorial clock would *actually* work more slowly than a polar clock, not merely *appear* to work more slowly, and if that deduction is "denying the central principle of relativity" as Maddox suggested, then that is clear evidence in support of the presence of an inconsistency in the theory.

In the passage quoted above, Maddox implied that the theory only requires one clock to *appear* to run slow from the other's point of view; it is therefore difficult to know what is meant by the following reference to Dingle in the penultimate sentence of the article: "And is there any hope that he will now be satisfied with the demonstration that *moving clocks run at different speeds from clocks at rest* ..." [Italics mine]. Clearly, if a moving clock runs at a different speed from a clock at rest, it must run either faster or slower; the writer of the editorial article is therefore "denying the central principle of relativity" in exactly the same sense as that in which he accuses Dingle of denying it.

Regarding Einstein's statement that a clock at the equator would work more slowly than a clock at a pole, the article has this to say:

> It seems now to be accepted that Einstein's original argument was uncharacteristically loose. The point of the illustration is that a clock at the pole of rotation may be taken to be in an inertial frame which is nearly (but not quite) properly defined by the direction of the Earth's motion around the Sun. The clock at the equator is in another. Einstein's lack of clarity concerns the inertial frame of the observer of the two clocks.

It is difficult to know what all this means, and it seems unkind to Einstein that the author of such a vague statement accuses *him* of looseness of statement and lack of clarity. If the writer of the above statement is suggesting that the equatorial clock does not *really* work slower than the polar one but only *appears* to some observer to do so, then he must reject Einstein's prediction that a clock that goes around in a closed path must actually show a different reading from one that stayed behind. It is interesting to compare the above quotation with what other reviewers of Dingle's book have written; their comments on the equatorial and polar clocks are discussed in Chapter 9.

One of the most interesting features of the leading article we are discussing (Anon., 1972) is the way in which it handles Dingle's reference to the other leading article, the one that was promised but never published. I discuss this part of the article in detail later in this chapter, after discussing Ziman's review.

Professor J.M. Ziman (1925-2005) was professor of Theoretical Physics at Bristol University. He developed interests in the social aspects of science and published several books on that subject. According to his Royal Society obituary (Berry and Nye, 2006), "on two occasions he was involved in negotiations that might have led to his becoming Editor of *Nature*." H.F. Judson (2004) described him as "an English physicist turned sociologist of science, a pugnacious little man with a distinctly original slant." His review of *Science at the Crossroads* was published in *Nature* under the title "Science in an Eccentric Mirror." (Ziman, 1973).

There are many features of Ziman's review that are worthy of study. For example, after quoting Dingle's question (as I have quoted it in Chapter 5, up to the words "How is the slower-working clock distinguished?"), Ziman wrote "This is a perfectly reasonable question to which science should indeed give an answer." Later in the review he wrote: "To distinguish the slower-working clock (as demanded by Professor Dingle) one must use the theory of *general* relativity, which takes account of accelerations." He made the following reply to the question: "In fact, the answer to Dingle's 'question' is simple: the fastest working clock between any two events is one that travels between them by free fall."

It is obvious from the above quotations that Ziman's statement is not a valid answer to Dingle's question. It is clear from the wording of Dingle's Question that the answer must be provided from the special theory, not the general theory. It is equally clear that Ziman's statement fails to answer the part of the question referring to the polar and equatorial clocks, which Ziman did not include in his quotation of the question. Since the answer is worded in terms of clocks that travel between two events, it appears that Ziman was still thinking in terms of the clock paradox problem in which clocks start together and then come together again later.

Since Dingle's Question explicitly asked about the special theory and only the special theory, as it was described in Einstein's 1905 paper, any valid answer to Dingle's question must have the property that it would have been a valid answer if the question had been asked in 1905, just after Einstein's paper was published and before the general theory appeared.

Ziman also described Dingle as a crank. He did not do so directly, but referred in his review to "another crank, Alfred Wegener," thereby insulting both men.

Soon after Ziman's review appeared, another scientist (Ellis, 1973) described Ziman's review as "admirable" but wrote that "Ziman invokes general relativity at a stage when it is not really needed." He went on to say: "In special relativity, just as in general relativity, the answer to Professor Dingle's 'question' is: the fastest working clock between any two events is one that travels between them by free fall."

We see that Ziman wrote that the answer to Dingle's question requires the use of general relativity, whereas Ellis wrote that it does not require it: that is a clear contradiction, on an important property of a theory. In other words, Ellis contradicted what Ziman had written in his review, but yet described Ziman's review as "admirable." If Ziman and Ellis had been interested in the search for the truth of the matter, they might have been expected to try to resolve that contradiction, but there is no record of any further discussion between them on that subject. The inevitable question arises: if those scientists were unable to see that their arguments contradicted each other, why should we believe them when they claim to know that there is no contradiction in the special theory?

Dingle (1973a) himself published a criticism of the Ziman-Ellis answer, saying "Neither of the events need be at either of the clocks concerned, so the statement, 'the fastest working clock between any two events is one that travels between them by free fall', is futile." The same defect of Ellis's answer was pointed out by H. L. Armstrong (1973) in the July 6 issue. Armstrong also went on to make the following very pertinent comment:

> Dingle remarked that everyone who answered his question had a different answer. Maybe the time has come for all of those who want to answer to get together and to come up with one official answer. Otherwise, the plain man, when he

> hears of this matter, may exercise his right to remark that when the experts disagree they cannot all be right, but they can all be wrong.

I would like to comment further on Ellis's claim that the answer given by special relativity to Dingle's question is that "the fastest working clock between any two events is one that travels between them by free fall." Suppose that there is a perfectly spherical, non-rotating planet without an atmosphere, and suppose that there is a clock carried in a circular orbit having a diameter just greater than that of the planet, so that once in each orbit the clock passes just over a stationary clock on the surface of the planet. According to Ellis's answer the orbiting clock, which is obviously in free fall, would record a *greater* reading for each interval between the close encounters of the two clocks; but according to the special theory a clock that goes round a circular path and returns to another clock would show a *smaller* reading for each interval.

Like Maddox, Ziman also fell into the trap of confusing the clock-paradox controversy with Dingle's claim that there is a contradiction in the theory. For example, commenting on Dingle's claim that there has been an inadequate public reply to his objections, Ziman wrote: "But here, again, he is grossly unfair. The clock paradox, and its resolution, was discussed in detail by Einstein himself, and by many later scholars." Since Einstein was dead before Dingle ever claimed that there was a contradiction in the theory, it is ridiculous to suggest that Einstein refuted Dingle's claim. After citing some books (none of which discuss the claimed contradiction to any significant extent) and mentioning how thoroughly they had been studied, Ziman said that this was as much of an answer as Dingle could reasonably expect, and then went on to say: "The fact that he, one man in a thousand, thinks differently is scarcely a major flaw in the scientific consensus."

Ziman's statement about the "many scholars" who are supposed to have addressed Dingle's arguments is a very weak attempt to claim that Dingle had been refuted. It would be much more convincing to cite one knock-down argument. Also, his idea of scientific consensus is rather flexible. Here is what he wrote about it elsewhere (Ziman, 1968):

> The '5% significance level' almost implies that 1 in 20 of one's colleagues is entitled to disbelieve one! What we must seek is *overwhelming* evidence that will persuade *everybody*—one doubt in 10,000 might be too many.

In the last paragraph of his review, Ziman described Dingle's book as "dishonest." Dingle told me some time later about the subsequent developments. He said that Maddox had subsequently refused to publish a letter from him asking for substantiation or withdrawal of Ziman's assertion of the "dishonest" character of his book, and had instead proposed publishing a brief letter from Ziman repeating the assertion but offering no evidence. Dingle therefore decided to take legal action, as a result of which he was able to have an apology from *Nature* and Professor Ziman published in the June 8 1973 issue (p. 315). That apology may serve as a confirmation that Dingle's narrative is factually accurate, and it is in the truth of its factual statements that, according to a letter from Dingle in the same issue (Dingle, 1973b), the whole significance of his book lies.

I shall now discuss the reference in the *Nature* editorial article (Anon., 1972) to the editorial article that was promised but never published. Here is what the published article says about the one that was not published:

> Professor Dingle goes on to complain that a promised leading article rounding off the correspondence has never appeared, apparently oblivious of the way in which his own scorn for prospective contestants and his promises to "bring discredit on the journal" may have discouraged the judicious summing-up for which he asked.

This quotation gives the impression that Dingle had asked for the leading article to be written, and also implies that, because of his alleged promise to "bring discredit on the journal," he was himself responsible for its non-appearance. I thought it was uncharacteristic of Dingle to promise to "bring discredit on the journal," and I asked him, in a letter dated March 7, 1974, whether those words were in fact a quotation from something he had written. He replied in a letter dated 1974 March 14, in the following words:

> The *Nature* editorial you mention was such a disgusting piece of chicanery throughout that I thought it beneath notice and ignored it; if I had sent a reply I am sure Maddox would not have published it. The item you mention is a typical example. In my long correspondence with him I repeatedly begged him not to bring discredit on *Nature*, as I had a real affection for the journal through my part in writing Lockyer's Life and in working in close association with Gregory for many years. Here is what I take to be the original of the passage which he puts in quotes: it occurs in a let-

> ter of May 20, 1969: "...It is now about nine months since I sent you my reply to Synge; it remains unpublished, though Landsberg's irrelevancies have appeared in the meantime. Naturally I have had to take other steps, which are still proceeding, and it is becoming increasingly difficult to present *Nature*'s attitude without reflecting discredit on the journal. I repeat that I do not wish to do this, but I must get the truth brought out, whatever it involves, and you are leaving me no alternative..." You can form your own judgment of the honesty of his quotation, and make whatever use you like of this trick.

Since, as is described on pages 89-90 of *Science at the Crossroads*, Maddox had written to Dingle on 24 November 1969 promising "to write a long leader summarising the position," Dingle's earlier letter of May 20 1969 could not have been the reason for the leader's non-appearance; nor had Dingle asked for the article as Maddox said he had.

Later in 1974 I wrote an article called "The Dingle Affair: An Unresolved Scientific Controversy," which I submitted to *Science Forum*, a Canadian journal of science and technology (now defunct), and which was to have been published in February 1975. In it I mentioned the *Nature* editorial (Anon., 1972), with its accusation that Dingle had promised to "bring discredit on the journal," and pointed out that Dingle's letter to Maddox in May 1969 could not have been a valid reason for the non-appearance of the leading article that Maddox had promised in November 1969. When the Editor of *Science Forum* sent a copy of my article to Mr. Maddox for his comments, Maddox was able to raise doubts in his mind about the authenticity of my article, and told the Editor that the article might be damaging to him, with the result that the Editor of *Science Forum* withdrew my article from publication.

Maddox, who was no longer editor of *Nature*, had complained that I had not asked for his side of the story. Although I thought that he had had ample opportunity when he was editor to write whatever he wanted, I wrote to him and asked him for his side of the story. He replied on 13 January 1975, saying that he was not prepared to give me a general statement about what Dingle had said on page 89 of *Science at the Crossroads*, but offering to answer specific questions. There was one quite striking passage in his letter, in the following words:

> What eventually happened was that Dingle wrote a letter which seemed to me at the time to be even more full of

> somewhat unrealistic threats of what he would do if the promised leading article did not turn up soon. I don't have a copy of that letter but I am sure Dingle has.

When I told Dingle what Maddox had written, he realized that the editorial article had been rather defamatory of him, and he decided to take some action himself. As a result of that action *Nature* published a letter from him (Dingle, 1975a), which described in some detail the correspondence between him and Maddox. Maddox published a reply in the same issue (Maddox, 1975), in which he referred to the timing of the letters as follows:

> On one small point, whose relevance to his complaint is debatable, Dingle is right. His promise to "bring discredit" on *Nature* was made before and not after I rashly volunteered to write a leading article explaining why his argument is false.

In my article that was withdrawn from publication in *Science Forum*, the fact that Dingle's letter expressing reluctance to reflect discredit on *Nature* was written before Maddox promised to write the leading article was the only item of factual information that was not already in the public domain. Although it was then so important to Maddox that he told the Editor of *Science Forum* that my article would be damaging to him, when he was forced to admit that fact publicly he dismissed it as a small point whose relevance was debatable. It is also interesting to note that his reply again mentioned Dingle's promise to "bring discredit" on *Nature*, even though it had by then been established that what Dingle had written was a plea to the Editor of *Nature* not to make it necessary for him to *reflect*, not *bring*, discredit; putting the words "bring discredit" in quotation marks in the letter merely makes that a quotation from the original misquotation in the editorial article (Anon., 1972). After quoting a sentence from a letter from Dingle to him on April 6, 1971, Maddox wrote that "the correspondence had become offensive and repugnant." He also wrote, among other things, that Dingle's error was primitive, and that "I note that he does not refer to the simple statement of the reasons why he is mistaken presented in the leading article of which he now complains." As I have already pointed out, that leading article (Anon., 1972) was a very poor attempt to answer Dingle's arguments.

Dingle replied to Maddox's comments in a further published letter (Dingle, 1975b), in which he quoted extensively from his letter that Maddoxhad mentioned, dated April 6, 1971, and added the following comment: "Your readers can judge whether the 'offensiveness and repugnancy' of this letter adequately excuse the succession of broken promises a year earlier." Shortly after that letter (Dingle, 1975b) appeared, Dingle told me how his lawyer had helped to ensure its publication "after much resistance too detailed to describe briefly." I have never seen any evidence that Dingle ever promised to bring discredit on the journal, as Maddox had accused him of doing.

I described some of the above-mentioned events in *The Relativity Question*, and it is interesting to read the comments of Professor Petr Beckmann (1924-1993) in his review of that book in the first issue of the journal *Galilean Electrodynamics* (Beckmann 1990):

> The author lets the facts speak for themselves and makes no accusations. Nor does he have to: the dishonesty of *Nature* editor John Maddox, for example, is impossible to miss, regardless of any scientific issues.

Maddox also used other *Nature* editorial articles to criticize the critics of special relativity. Consider the following statement, which is the first sentence of his editorial article "More tests of special relativity" (Maddox, 1986) :

> Given the zeal with which those who disbelieve the special theory of relativity go about their business, it is a comfort that experimental tests of relativity are also flourishing.

Although Maddox did not refer to any particular group of disbelievers, it is reasonable to assume that he included those who claim that the theory is self-contradictory. But, as I have already pointed out, there is no experimental evidence that can refute that claim, because a theory that is self-contradictory can, by valid logical rules of inference, be used to predict any experimental result whatever. I discuss this subject in more detail in Chapters 8 and 11. The last sentence of Maddox's article is also interesting; it reads as follows:

> Much has been done already, but further efforts are needed, and not merely to keep the army of sceptics of the special theory at bay.

In both the above two quotations, Maddox seems to be implying that disbelief in special relativity is a bad thing and that it is desirable for *Nature* (and Nature) to provide support for relativity. I think that it is well outside the scope of an editor of a scientific journal to make such value judgments. Perhaps it may be appropriate to apply to Maddox what Henry John Stephen Smith said of the first editor of *Nature*, Sir Norman Lockyer, that "he sometimes forgets that he is only the editor and not the author of Nature." (Glaisher, 1894).

Professor Ziman appeared again in the story in 1977. In April of that year Herbert Dingle's collaborator Mark Haymon wrote a letter to Ziman, pointing out respectfully that his published answer to Dingle's question was not satisfactory, and asked him, since Ziman had written that science should indeed give an answer to the question, who should be expected to give the answer on behalf of science. Ziman replied that Dingle's views on the clock paradox had been given adequate attention in a variety of publications, and that he could not demand more than that. It seemed surprising that he still harped on the clock paradox, since Dingle had made it very clear in his book that he was not still debating the clock paradox. Haymon replied to Ziman, pointing out that the discussion was not about the clock paradox, that no one had answered Dingle's crucial question, and asked him to stop evading the issue and answer Dingle's question.

Ziman replied that, as he had indicated in his review, many able scientists had given time and trouble to Dingle's question and had presented their views on it, and that those views could be found in a number of widely published documents. His reply was completely evasive: it is a very easy thing to say that many people have presented their views about Dingle's criticisms, without identifying any of them; it is much more difficult to identify *one* rebuttal as the crucial knock-down argument. As I pointed out in my article *Problems in Special Relativity* (Appendix II) the scientists who claim to refute Dingle cannot all be right, and the possibility exists that they may all be wrong.

In his second letter to Ziman, Haymon referred to a recent letter in *The Times*, written by Ziman as Chairman of the Council for Science and Society (Ziman, 1977), in which he had written that "It is the particular concern of the Council for Science and Society to study the social consequences of science and technology." Dingle and Haymon decided to describe the problem to the Council, in the hope that it might help to draw attention to the unsatisfactory situation, and they sent a joint letter to each Member of the Council, pointing out the unsatisfactory state of affairs in that no proper answer had been given to Dingle's question about the special theory, the question that Ziman had described as "a perfectly reasonable question to which science should indeed give an answer." They included with their letter copies of the correspondence between Haymon and Ziman. I reproduced Haymon's letters to Ziman, and the joint letter to the Members of the Council, in full in *The Relativity Question* (McCausland, 1988); I did not reproduce Ziman's replies to Haymon, because Ziman had refused to give me permission to publish them.

The response of the Council for Science and Society was transmitted by Professor Ziman in a letter dated 11th October 1977, addressed to Mr. Haymon. The letter, which Professor Ziman has also refused to give me permission to publish, stated that the correspondence had been discussed at a meeting of the council in June, that it had been decided to make no response in the name of the Council, though individual members might reply as they wished, and that Professor Ziman himself did not propose to make any further response to Mr. Haymon's communications. The letter did not explain why the result of a decision made in June was not transmitted to Mr. Haymon until October. Six members of the Council had in fact written to Haymon and Dingle, who had naturally assumed that those members were writing with a view to making a contribution to the discussion, whereas the decision had already been made in June. One surprising feature of Ziman's letter was that, although it reported the action (or inaction) of the Council in response to the joint Dingle/Haymon letter, it was addressed to Haymon alone and did not even mention Dingle.

Although I had asked Ziman for his permission to publish his letters to Mark Haymon, I do not think I need his permission to publish his letter to me, and I quote it below. I had written to him to say that I was working on a book that would be a sequel to *Science at the Crossroads*, and that, as was done in that book, I intended to quote the exact wording of various letters; I specifically asked for his permission to publish the three letters he had written to Mark Haymon. He replied as follows, in a letter dated 30th August 1979:

> Naturally, I should like to help in any serious academic project describing the relationship between Professor Herbert Dingle and the rest of the scientific world. But as I made quite clear in my review of *Science at the Crossroads* (herewith enclosed) his view of the issue was gravely distorted and it would be most important that any sequel to this book should get the matter quite straight. Indeed, since that re-

> view was written with just that purpose I hope that you will republish it—not, of course, selectively but in full so that any reader of your book would have a very clear idea of the way in which almost all physicists, with any technical knowledge of the subject, actually felt about Dingle's ideas.
>
> On the other hand, the letters exchanged with Mr. Haymon, whether by myself or by other members of the Council for Science and Society, were not intended to publication and I do not believe that any good purpose would be served by reproducing that correspondence. I am not willing, therefore, to give my permission for publication of these letters in whole or in part.
>
> If you want to come and talk to me about this I shall be happy to discuss it with you, but I hope you will forgive my unwillingness to be drawn into this dotty affair, now that the poor old chap is dead.

I found it interesting that his enclosed typescript of his *Nature* review (Ziman, 1973) still included his accusation of dishonesty and his misspelling of McCrea's name; also that he seemed to think that he had the right to refuse me permission to publish letters written by other Members of the Council, two of whom had given me their own permission and whose letters I reproduced in *The Relativity Question*. I think it is unfortunate that the Council for Science and Society did nothing to help Herbert Dingle and Mark Haymon in their attempts to get an answer to Dingle's question, which Ziman had agreed was "a perfectly reasonable question to which science should indeed give an answer" but had not answered.

I think it is a considerable stretch of the imagination on Ziman's part to suggest that his review represented "the way in which almost all physicists, with any technical knowledge of the subject, actually felt about Dingle's ideas." Did almost all physicists really think that it was valid to invoke general relativity to answer a question that was explicitly asked about special relativity?

Not only did Maddox and Ziman make it difficult for Herbert Dingle to present his case to the scientific community, they also made it difficult for me to describe how they did it.

PART THREE

Continuing the Adventure

This part describes some of my published papers about relativity in various journals. Most of the these were published after Herbert Dingle's death, and several of them are included as appendices at the end of this book. Although most of my published papers were about the special theory, I also published a paper about the findings of the 1919 eclipse expeditions that tested the general theory. In Chapter 7 I describe some of my papers that were the subjects of published comments by others, and add my own comments in reply. In Chapter 8 I describe the debate between Jack Good and myself in the pages of *Physics Essays*, arising from my paper "An Inconsistency in Special Relativity" which is reproduced in Appendix IV.

Chapter 7

My Published Criticisms of Relativity

> With hindsight we now know that all predictions of general relativity had to await the 1960s and decisive technological innovations before they could be tested accurately on a firm *quantitative* basis. All earlier allegedly quantitative confirmations turned out to be unsatisfactory, sometimes even illusory.
>
> Klaus Hentschel (1992, 598).

> Einstein's first paper on special relativity, published in 1905, is axiomatic in structure; the whole edifice is erected on new first principles. It is so perfectly written that what remained to be done ever after was to work out further consequences of the Einstein postulates. Not one word in the paper needs to be changed in the light of later developments.
>
> Abraham Pais (1994).

After I became actively interested in the debate on special relativity, I tried to publish some papers critical of the theory. My first publication was a short note in the *Proceedings of the Institution of Electrical Engineers* (McCausland, 1972), which led to a short, inconclusive discussion. I started my main correspondence with Herbert Dingle by sending him a copy of that note, but that was the only critical paper that I was able to publish in the scientific literature during his lifetime. I tried to publish papers in various other journals, and was soon struck by the dogmatism and irrationality that I encountered from editors and referees. So, when I was the editor of the *Canadian Electrical Engineering Journal*, I thought I would take the opportunity to publish a correspondence item (McCausland, 1980a) pointing out some of the confusion that existed among the various replies that had been made to Dingle's criticisms. I sent copies of the article to the people whose replies I had criticized, but without any reaction from any of them.

Herbert Dingle had sent me the manuscript of his last scientific paper, on the twin paradox, and after his death I succeeded in having it published in *Wireless World* (Dingle, 1980). To accompany his paper I wrote a short biographical note and a reply, on his behalf, to an article (Davies, 1980) that had criticized him. As mentioned earlier, I later published another article in the same journal (McCausland, 1983b) with the title "Problems in Special Relativity." In that article I pointed out yet more confusion and inconsistencies in the arguments that had been used to defend special relativity from Dingle's criticisms. Most of these arguments had been published either in the debate in *The Listener* in 1971 or in reviews of *Science at the Crossroads*. Although that article is now more than a quarter century old, everything in it is still valid and the confusion and inconsistencies that I documented have been neither acknowledged nor corrected. There is not a word of that article that I wish to retract today, and I think it is one of my best short statements of the problems involved in debating the theory at that time. The article is reproduced in Appendix II.

I think it is useful to record one scientist's reaction to that article, which referred to two inconsistent statements made in the correspondence that I mentioned in Chapter 1, in *The Listener* in 1971. As mentioned earlier, that correspondence started when Professor J. Taylor claimed in an article (Taylor, 1971b) that the results of the well-known experiment of Hafele and Keating, which had then been recently performed, supported Einstein's special theory. Professor Dingle published a letter in the 18 November issue, answering Taylor's article, and another scientist, Professor M. A. Jaswon, published a letter in the 25 November issue which disagreed with some of Dingle's points but which agreed with Dingle that the experiment in question had "no relevance whatever for the special theory." Although that statement was directly contrary to Taylor's claim, Taylor later published another letter in the 9 December issue which ignored Jaswon's statement but continued to criticize Dingle.

Although I did not mention it in my article, that was the letter that stimulated Bernard Levin into describing in *The Times* how Taylor had evaded Dingle's question that called for a plain yes or no, and that was the letter in which Taylor insulted Dingle by telling him that "Those that can, create; those that can't, criticise."

I sent copies of my article to several eminent professors whose writings I had criticized in it, and Professor Taylor was kind enough to write to me about it, in a letter dated 8 November 1983. He wrote that he might not have known about Professor Jaswon's letter when he wrote his letter that was published in the December 9 issue of *The Listener*. That is extremely unlikely, to put it mildly, since Jaswon's letter appeared in the same issue as one of Taylor's own letters (25 November). Taylor would certainly have looked for his own letter in that issue, and it would have been hard to avoid seeing Jaswon's letter, especially as it appeared on the same page as the ending of his own letter.

Professor Taylor also told me that he felt that I had a good point that there was indeed something to be cleared up about the issue. His initial feeling was that he was not right in saying that the Hafele-Keating experiment justified special relativity but that Dingle was still wrong in his claim of an inconsistency. He agreed that one of the other reviewers of *Science at the Crossroads* whom I criticized in the article had been rather unconvincing in what he wrote, but said that it is convincing enough when properly explained. Regarding the polar and equatorial clocks, he wrote that the only proper explanation was in terms of general relativity; this obviously conceded one of Dingle's main points: that Einstein did *not* have the right to claim that the slowing of the polar clock followed from the special theory.

This letter was very significant to me, in that it was the first letter I had ever received from a relativist admitting that there was any flaw in the relativists' case. Unfortunately, however, Professor Taylor rejected a suggestion I made some months later, that he publish some clarification of the issue, and also refused my request for permission to publish what he had written to me in his letter. He was too busy with his commitments to his research programme, his students, his Department and his College. Of course, these commitments had not interfered with his criticizing and insulting Herbert Dingle when he was alive. By now Taylor has reached retirement age, so he might be expected to have time to look into the matter again, but to the best of my knowledge he has done nothing more about it. In retrospect, I think I was too scrupulous in asking his permission to quote what he wrote in his letter: I should have simply quoted publicly what he wrote and hoped that he would reply publicly. It might have drawn some attention to the subject and helped to settle the matter earlier. Perhaps I am still too scrupulous, but I will continue to respect his refusal of permission to quote his letter. I still have all the correspondence, however, and in reading it I am struck by the contrasts: between the eagerness of his pursuit of the heretic in 1971 and his reluctance to become involved in 1984; between his aggressiveness towards Dingle in 1971 and his timid unwillingness even to be quoted in 1984. Although he was very busy in 1984, that would not have kept him from giving me permission to quote what he had written.

For some years after the publication of that article I worked on a book that would tell some of the story of my collaboration with Herbert Dingle and that would document the correspondence mentioned earlier, between Herbert Dingle and Mark Haymon and various people and organizations, copies of which Dingle and Haymon had sent to me with the intention that they would eventually be published. I was unable to find a publisher for the book, so I eventually published it privately under the title *The Relativity Question* (McCausland, 1988). The word "Question" referred to the question that was the central feature of *Science at the Crossroads*, and my book was intended as a kind of sequel to that book.

As mentioned above, *The Relativity Question* documented some of the correspondence between Dingle and Haymon and various people and organizations. As I pointed out in the book, the correspondence that I published there is somewhat one-sided because some of the other correspondents refused my requests for permission to publish their letters to Dingle and Haymon. Although some of the correspondents who refused their permission are now dead, I shall respect their refusal.

My attempts to publish criticisms of special relativity became more successful after I made the acquaintance by correspondence of Tom Phipps (Dr. Thomas E. Phipps, Jr.). He wrote to me on 23 March 1987 to compliment me on the above-mentioned articles "Science on the defensive" and "Problems in special relativity," and as a result I have been in correspondence with him for more than twenty years; he and I have often exchanged copies of drafts of our publications, and when *The Relativity Question* was published in 1988 I sent him a copy. Then in February 1989 he told me about the journal *Physics Essays*, which had then been recently established, and suggested that I might wish to contribute a paper to it. Since he had also complimented me on one particular argument in *The Relativity Question*, I decided to write a paper based on that argument, and it was published in *Physics Essays* (McCausland, 1990). This article is important because it later formed the start of my published debate with Professor I. J. Good, and it is reproduced in Appendix IV. Since that debate represents an important part of my story, I describe it in detail in the next chapter.

As I mentioned in Chapter 5, another paper that I published in *Physics Essays* was "On Faraday's Ray-vibrations and the Postulates on the Velocity of Light" (McCausland, 1992) which is reprinted in Appendix III.

Another journal in which I was able to publish papers was the *Journal of Scientific Exploration*. I had become aware of the Society for Scientific Exploration, which publishes the journal, after I became acquainted with Professor Henry Bauer, of Virginia Polytechnic Institute and State University, in March 1985. I had written to him to comment on his book *Beyond Velikovsky* (Bauer, 1984), in which he had expressed a sympathetic point of view about Dingle's *Science at the Crossroads*. As I describe in more detail in the next chapter, it was through Henry Bauer that I got in touch with Jack Good, who was a very important participant in my scientific adventure.

One paper that I published in the *Journal of Scientific Exploration* was "Anomalies in the History of Relativity" (McCausland, 1999). This paper, which is reprinted in Appendix V, arose from my study of the history of relativity, and describes the results of the British eclipse expeditions to Principe and Sobral in 1919 that made the measurements that were taken to support Einstein's general relativity theory. One purpose of the observations was to distinguish between a smaller deflection of light that would have been compatible with Newton's theory, and a larger deflection predicted by Einstein's. I pointed out some of the shortcomings of the measurements and reviewed other papers showing that the results were not really decisive.

In my article I also described the joint meeting of the Royal Society and the Royal Astronomical Society on 6 November 1919, at which the results of the eclipse expeditions were announced. The significance attached to the meeting might be judged by Pais's statement, in his book *"Subtle is the Lord ..."* (Pais, 1982, 305) to the effect that Einstein was canonized at the meeting. Although Pais used *The New Catholic Encyclopedia* as his reference for the process of canonization, his comparison was inaccurate. He conflated two quite different meetings: the meeting of the Congregation of Rites, at which a candidate is considered for canonization, and the event at which the Pope formally pronounces the canonization. He referred to the presence of a devil's advocate, a person whose function in the canonization process was well described by John Stuart Mill (1874, 40) in his book *On Liberty*:

> The most intolerant of churches, the Roman Catholic Church, even at the canonization of a saint, admits, and listens patiently to, a 'devil's advocate.' The holiest of men, it appears, cannot be admitted to posthumous honours, until all that the devil could say against him is known and weighed. If even the Newtonian philosophy were not permitted to be questioned, mankind could not feel as complete assurance of its truth as they now do.

If Mill had been writing a century later, he would probably have written "Einsteinian" instead of "Newtonian"; it is certainly equally appropriate. It is obviously imperative that the arguments of the Devil's advocate be heard *before* canonization is pronounced, and Pais's account gives the impression that the criticisms of his devil's advocate, Ludwik Silberstein, had been heard before the results were endorsed by the President of the Royal Society. However, the generally accepted account of the meeting (Anon., 1919) shows that Thomson had, to use Pais's words, "pronounced the canonization" before Silberstein had had a chance to speak.

Einstein's rapid rise to enormous fame was a direct result of the announcement that the measurements made during the eclipse expeditions supported his general theory. As a result of the enormous prestige that he has been accorded ever since then, both scientists and non-scientists find it very difficult to believe that there can be any flaw in either general relativity or special relativity, and critics of the special theory tend to be dismissed as crackpots. That was why I ended my article with the sentence: "In the author's opinion, the confident announcement of the decisive confirmation of Einstein's general theory in November 1919 was not a triumph of science, as it is often portrayed, but one of the most unfortunate events in the history of 20th-Century science." I should admit that is a rather strong statement, but I believe it is justified by the fact that the grossly excessive adulation and hero-worship of Albert Einstein, which still makes the scientific world unwilling to examine his theory critically, arose directly from the confidence of the announcement that was made at the famous meeting in 1919.

The most critical comment on my article that I have seen was by Sahotra Sarkar (2007) in his book *Doubting Darwin?* He described me on page 155 as "an enthusiast of some now justly forgotten alternative theory of Paul Gerber." I had correctly pointed out that Gerber had used a delayed-propagation theory to derive and publish the same expression as Einstein, for the variation of the perihelion of Mercury, before Einstein had derived it, and I had suggested that the British scientists might not have known of Gerber's work when they were considering the merits of the general relativity theory in 1919. After quoting the last sentence of my article, Sarkar continued: "According to McCausland, the acceptance of relativity closed people's minds, preventing other theories from being heard. The rhetoric is identical to that of ID [Intelligent Design] creationists except that their target is the theory of evolution." Even though he did not show any factual error in my paper, he wrote that "few scientists bother with dissent of as low credibility as McCausland's, and ID is no different in this respect." In the endnote attached to his quotation of my last sentence, he gave the citation for my article and then wrote: "This was published in a peer-reviewed journal but the peers seem largely to consist of those who believe in telepathy, reincarnation, folk reports of animals unknown to zoology, and so on." This last comment seems to me to be about a third or fourth level *ad hominem* argument: it is not based on my article, or on me, or even on the referees of my paper, but on what seem to be the beliefs of the referees of papers on other subjects in the journal. Sarkar is a Professor of Philosophy at the University of Texas at Austin; I hope that he does not encourage his students to use *ad hominem* arguments.

It is interesting that Professor Sarkar commented on the scope of papers published in the *Journal of Scientific Exploration*. Although there are indeed many articles on what might be called fringe areas of science, many of which stretch one's belief, one article that aroused indignation among its readers was "The Relative Motion of the Earth and the Ether Detected" (Gift, 2006). One reader (Osborne, 2006) even wondered whether the journal had been the victim of a hoax. The story is told in two editorial articles (Bauer, 2006, 2007). The very strong reaction to the paper exposed a serious dilemma for the Journal, which was succinctly described by a reviewer giving the following advice to the editor:

> The whole problem of delving into the fundamental postulate of relativity theory is kind of explosive. Is it a subject that SSE wishes to get into? If so, you must get into it very deeply.

After considering the problem carefully, the editor declared that he intended thenceforth "to reject relativity-dissenting manuscripts without formal review." (Bauer, 2007) Although in the circumstances I sympathize with his decision, I think it is revealing that the *explosiveness* of the response to the critical article had such an influence on the decision.

Another interesting (and unintentionally amusing) comment about Gerber's work was made by Tim Maudlin (1994, 28):

> Gerber's work was unfortunately seized upon by rapid anti-relativists and has fallen into disrepute; the notion of a delayed-propagation Newtonian gravitational theory, though, is natural enough to warrant study.

By the time the second edition of Maudlin's book appeared in 2002 the anti-relativists were rabid. His suggestion that a delayed-propagation gravitational theory is "natural enough to warrant study" might be compared with my own published comment in the article: "Perhaps it might be interesting to make an objective reappraisal of Gerber's work." Presumably that is the phrase that, according to Sarkar, showed me as "an enthusiast of some now justly forgotten alternative theory of Paul Gerber."

Another book that mentioned my paper is by Matthew Stanley (2007). His book discussed in detail the handling of the eclipse observations, and concluded that they were fairly handled. An endnote on pages 268-9 included the following: "The mythology of a conspiracy to support Einstein is widespread and influential. For examples of such criticisms outside science studies see ... [citations of three papers, including mine]." I did not suggest in the paper, or in anything else I wrote, that there was a *conspiracy* to support Einstein.

Much nonsense has been written about the bending of light shown by the eclipse observations. For example, Lee Smolin (2006, 43) wrote, referring to the bending of light that was studied by Eddington, "This effect was observed during a total solar eclipse, which made it possible to see, near the rim of the occluded sun, the light from stars that were actually directly behind the sun. Had the sun's gravity not bent their light, these stars would not have been visible." That statement is completely ridiculous: the bending of light was extremely small and certainly did not allow stars behind the sun to become visible. Another author who claimed that the telescopes of the 1919 expeditions detected stars that were behind the sun was B. L. Silver (1998, 431). However, those statements are modest compared with Figure 2.9 of Stephen Hawking's famous book *A Brief History of Time* (Hawking, 1988) which shows so much bending of light that a star that is actually to the left of the sun appears to be on the right! It seems surprising that the diagram survived unchanged up to its inclusion in "the updated and expanded tenth anniversary edition" of the book.

In any case, the canonization that Pais described is modest beside some other comparisons of Einstein's status to religious figures. For example, Paul Davies (1983) wrote in his book *God and the New Physics* that Albert Einstein "carries the same status in the scientific community as St. Paul does among Christians." An even higher status was accorded to him by Thomas Levenson (2003) in his book *Einstein in Berlin.* In his Prologue, titled "The Adoration," Levenson described the visit of Walther Nernst and Max Planck to Zurich in the summer of 1913 to try to persuade Einstein to go to Berlin, and compared their visit to the biblical story of the visit of the three kings to Bethlehem. Here is part of the comparison:

> The two Germans went to Zurich; the magi traveled to the manger. They brought both gifts and praise. But this time, they sought something in return: Albert Einstein, in Berlin.

Even those comparisons seem tame when compared with Corey S. Powell's (2002) book *God in the Equation.* On the jacket of this book we read:

> We are living at a turning point in human spirituality—akin to when Jesus or Buddha or Mohammed was alive—and Einstein is its prophet. That is the audacious, provocative, and fascinating argument Corey Powell makes with dazzling eloquence in this extraordinary book.

Powell made extravagant claims about what he called the Church of Einstein and the new faith of sci/religion that is associated with it. He wrote (p. 246) that "The Church of Einstein is more authoritative and comprehensive than ever. Its spiritual power has eclipsed that of the old-time religions."

Many scientists apparently worship in the Church of Einstein; for example we read on p. 245:

> At the American Astronomical Society, as well, there is more sci/religious zeal than the superficial sights and sounds indicate. Despite their reluctance to invoke overtly theological language, in their actions the current priests of sci/religion remain firmly committed to the Church of Einstein.

It has been pointed more than once (Lanczos, 1965; Stachel, 2002) that Einstein is both the greatest and the second-greatest modern scientist, the greatest because of relativity and the second greatest because of all his other accomplishments. Powell (2006) has gone even further and has written about "my three Einsteins, related but distinct aspects of the man, which I envision nested inside one another like Aristotle's heavenly spheres." Whether the godhead of Powell's Church of Einstein is a twosome or a threesome, it would be no more difficult to understand than the Doctrine of the Trinity, and many English worshippers in the Church could continue to describe their affiliation as C of E. Even though the spiritual power of the Church of Einstein has eclipsed that of the old-time religions, I do not expect to see the Collected Works of Albert Einstein supplant the Gideon Bible in hotel rooms throughout the world; nor will the citizens of this Brave New World decide to reset the calendar to zero at the date of Einstein's birth, or even the date of the famous eclipse that marked the start of the modern world—for one thing, the world's computers would never agree.

The deification of Einstein is so extravagant that some people almost seem to think that the universe was created following a design provided by Einstein. Einstein himself seems to have placed himself on a high level relative to the deity, as in his frequently quoted comment that he "would have had to pity our dear God" if the 1919 eclipse results had not supported his theory (Calaprice, 2011, 368). As Jaki (1989, 95) put it: Einstein "indulged in playing the role of scientific advisor to a Universe-God that had to obey his General Relativity and his Unified Field Theory."

I also published a rather light-hearted article "The Einstein Mystique" (McCausland, 2003) in which I discussed some of the reasons for Einstein's enormous prestige; the article is reproduced in Appendix VI. One of the phenomena I discussed was the eagerness of publishers to publish books including Einstein's name in the title and/or having his picture on the cover. An example that I find amusing but did not know of at the time is a book by John Waller in which he devoted one of thirteen sections to a critical discussion of Eddington's handling of the 1919 eclipse observations. When I first found this book in Britain it was a paperback with the title *Fabulous Science* (Waller, 2002a) and it had a picture on the cover of a flask of some evil-looking liquid. I later discovered that the same book was published with the title *Einstein's Luck* (Waller, 2002b) with a picture of Einstein on the cover! At least one online reviewer of the latter book wondered why Einstein's name was in the title, when so little of the book was about him. Waller was very highly critical of Eddington's handling of the eclipse results. Referring to the fact that some of the observations from the expedition to Sobral seemed, as Eddington himself wrote, to support the Newton result rather than the Einstein one, he wrote (page 58): "So serious were Eddington's manipulations that one strongly suspects that had the predictions of the rival theories been the reverse—Newton high, Einstein low—Eddington would have discarded his own photographs as too hazy and accepted with alacrity the Sobral party's astrographic pictures."

In spite of the fact that the results obtained by the Principe expedition were so poor, a commemorative plaque was unveiled at Principe on the 90th anniversary of the eclipse, in May 2009. Rothman's words (2003, 87) seem appropriate: "Only one result is sure: The tale that the 1919 eclipse expedition verified Einstein's theory will forever remain untarnished."

Another book that I did not know of when I wrote The Einstein Mystique was "Selections from *The Principle of Relativity*," Edited with Commentary by Stephen Hawking (2002a). This book is a collection of Einstein's papers, the same papers as appeared in the collection of papers originally published in English by Methuen (Lorentz *et al.*, 1923), together with a few pages of biography of Einstein. The book is described on the back cover as "this groundbreaking volume" which helps to make Einstein's work "accessible to a new generation of readers." As if anyone who was willing to take the trouble to read any of those papers could not gain access to a copy of the original collection, which contains more papers and in my opinion is presented in a more easily readable format. Hawking's volume is groundbreaking in the same way as an exhumation involves groundbreaking.

In addition to being published as a book, the Hawking collection was also incorporated into a larger collection (Hawking, 2002b; more than 1200 pages, and about 1.7 kilograms) which included also works by Copernicus, Galileo, Kepler, and Newton. There are several errors in mathematical expressions in Einstein's papers in the larger collection; most of these, but not all, were corrected in the smaller volume.

My article "The Einstein Mystique" appeared in 2003, before the massive celebrations of the hundredth anniversary of Einstein's Miraculous Year in 2005 led to an enormous increase in the number of books and articles about Einstein. Even some of Einstein's admirers thought it was rather excessive, one of whom wrote as follows (Janssen, 2006):

> The year has seen more than its share of forgettable talks, bad TV shows, and—worst of all, since they will continue to clutter bookcases for years to come—an avalanche of unimaginative books cashing in on the public's insatiable demand for *Einsteiniana*. Fifty years after his death, the world's most lovable egghead continues to be a superstar in the netherworld of top-grossing dead celebrities.

In that connection, I was amused by an article in a "Special Einstein Issue" of *Discover* magazine (Bodanis, 2004), in which he wrote that "Many of the pharmaceuticals we take, from statins to Viagra, were fabricated using techniques that draw on Einstein's 1905 paper on Brownian motion." Even though Einstein is already one of the "top-grossing dead celebrities," he might become really rich if his estate could claim royalty payments on sales of Viagra!

Some authors have published several books about Einstein. A notable example is Alice Calaprice, a longtime senior editor at Princeton University Press, with her four books of Einstein quotations published by that Press: *The Quotable Einstein*, 1996; then *The Expanded Quotable Einstein*, 2000, which is "enlarged and updated," even though Einstein had not made any more quotable statements in the meantime; then *The New Quotable Einstein*, 2005, which is "newly expanded and updated"; and finally (I hope) *The Ultimate Quotable Einstein*, 2011. She has also published *Dear Professor Einstein: Albert Einstein's Letters to and from Children*, 2002, and *An Einstein almanac*, 2005. I wonder whether she blushed all the way to the bank after publishing all those books.

Chapter 8

My Debate with Professor Good

> I like to resolve paradoxes. I recently wrote a paper on the kinematics of special relativity for that purpose. Herbert Dingle had claimed that there was a contradiction in the special theory and my paper answered briefly, and I hope clearly, all four of his arguments. (I also defended the kinematics of the special theory against an ingenious argument by Ian McCausland and two arguments by George Galeczki.) Dingle was misled by a remark of Einstein's that was not literally correct. I am convinced that the kinematics of the special theory is selfconsistent and can be disproved only experimentally, but there are still flat-earthers who think the theory is self-contradictory.
>
> I. J. Good, (Banks, 1996)

Professor I.J. (Jack) Good (1916-2009) was a very distinguished mathematician who was one of the senior scientists at Bletchley Park during World War II, where he was involved in deciphering German messages. He went on to a distinguished career both in England and in the United States, eventually becoming a Distinguished Professor of Statistics at Virginia Polytechnic Institute and State University in Blacksburg, Virginia. Interesting biographical and autobiographical information about him appears in two articles, both based on interviews: one by Christopher Evans (1979), and one by David L. Banks (1996), from which the epigraph quotation is taken. In one of his last publications (Good, 2008) he wrote that he had published more than 2000 technical papers.

Jack Good was a most interesting character, who was interested in many diverse fields of knowledge. He also had a very open mind about what might be called fringe areas of science, and was very tolerant of dissent. For example, in his interview with Evans he expressed his criticism of the way in which scientists had tried to silence Immanuel Velikovsky, and said that scientists had "behaved quite disgracefully" in the way they had treated him.

I think it is a great pity that Good did not engage in debate with Dingle on the subject of special relativity. He certainly was aware that Dingle believed that the special theory was untenable, since he acted as the General Editor of a book called *The Scientist Speculates: An Anthology of Partly-Baked Ideas* (Good, 1962), to which Dingle contributed an article "The Observation of Line Events" (pp. 337-348), in which he stated that "I have no doubt whatever that the special theory of relativity is no longer tenable." An Editorial footnote was appended to Dingle's article, presumably by Good himself, but it made no comment on that statement. Both Dingle and Good were highly argumentative people, and I believe that a debate between them on the subject of special relativity would have been an extremely rich contribution to science. My own debate with Good is a poor substitute for such an encounter, but I believe that it is interesting in its own right.

I made the acquaintance of Professor Good through his colleague at Virginia Polytechnic, Professor Henry Bauer, whom I mentioned in Chapter 7. In the course of my correspondence with Bauer I had sent him a copy of *The Relativity Question* (McCausland, 1988) and he had later shown this to Good. After reading some of that book, and some of Dingle's *Science at the Crossroads*, Good wrote to Bauer and me in May 1990, with critical comments on some of my arguments. Since he was obviously eager to continue discussion of the matter, I sent him in July 1990 a copy of my *Physics Essays* paper, (McCausland, 1990) which I mentioned in the previous chapter, and invited him to publish a reply. This paper was the first of several papers in the debate; the complete list of papers is given at the end of this chapter.

I met Good in person at the 1991 meeting of the Society for Scientific Exploration at the University of Virginia in Charlottesville. I was pleased to meet him and I enjoyed his company. Since we had agreed that the debate on special relativity should be carried on in writing and in public, we did not discuss the subject to any significant extent at the meeting; in fact his first paper in *Physics Essays* (which was in part a reply to my own paper), was in press at the time.

Although I do not think it is appropriate to discuss the debate in great detail here, since the interested reader can read it in the journal, I shall present some of the most relevant arguments. For brevity I shall designate the papers of our debate by numbers in square brackets referring to the list of references at the end of this chapter. The argument in my first paper [1] can be summed up very simply. I started in that paper by referring to Einstein's statement, in his original paper on special relativity, that a clock at the equator would work more slowly than an identical clock at one of the poles, and then stepped backwards in his argument to the case of a clock moving around a polygonal path. I considered a clock A at the centre of a square, and supposed that another clock B moved around the perimeter of the square at a very high constant speed. According to special relativity clock B would lose a definite amount of time in each complete tour around the square, and by symmetry it would lose a quarter of that amount while moving along each side. Hence B would work steadily more slowly than A while travelling along any one side. Now suppose that a third clock C moves at uniform speed along an infinite straight-line path in such a way as to travel alongside B while B is travelling along one side of the square. Then C would work at the same rate as B, steadily slower than A. But, by the principle of relativity, we can equally well show that A must work steadily more slowly than C since A and C are in uniform relative motion throughout, so there is an inconsistency.

Good's first answer to my criticism of the theory appeared in his paper "The Kinematics of Special Relativity, and Dingle's Fallacies," [2] which criticized my paper in addition to some of Dingle's arguments. His criticism of my paper was easy to answer, since he concentrated his argument on sudden changes or jumps in the *perceived rates* of clock B at the vertices of the square. Changes in the perceived rates would occur whether relativity is correct or not, and they were not relevant to my argument.

Good then embarked upon a series of articles under the general title of "The Self-Consistency of the Kinematics of Special Relativity." His articles were highly mathematical, and ranged widely over various subjects, with the result that much of their content was not relevant to my criticism; by contrast, I kept my own articles as short as possible, and used no mathematical equations whatever.

One of Good's main arguments against my thesis was that Einstein's statement about the polar and equatorial clocks was an error. He stated [4, 438] that the statement was "at least a lapse in exposition," and wrote that "it is possible that Einstein did make a mistake and *meant* what he literally said." In the "Conclusion" section of that paper he wrote that "in Einstein's exposition of the twin paradox, there is a sentence that, if taken literally, does indeed lead to a contradiction."

Since Einstein's statement about the polar and equatorial clocks is a central part of the debate, I will now quote it in full; in order to be able to refer to it simply and precisely. Since it is important to know the context of the statement, I shall also quote the previous sentence, and I shall use the labels E1 and E2 that I used to refer to these statements in the course of the published debate. The following two statements (without the labels) constitute a complete paragraph of Einstein's paper:

> E1. If we assume that the result proved for a polygonal line is also valid for a continuously curved line, we arrive at this result: If one of two synchronous clocks at A is moved in a closed curve with constant velocity until it returns to A, the journey lasting t seconds, then by the clock which has remained at rest the travelled clock on its arrival at A will be $(1/2)tv^2/c^2$ second slow.
>
> E2. Thence we conclude that a balance-clock at the equator must go more slowly, by a very small amount, than a precisely similar clock situated at one of the poles under otherwise identical conditions.

The next paper of Good's that is relevant to this discussion is [8]. This paper is perhaps the most striking of the whole debate, for various reasons. In that paper Good repeated his claim that E2 should not be taken literally, and gave it the name "Einstein's Slip" on page 253. He continued to say that the statement, if taken literally, did lead to a contradiction, and made the following very interesting comment, (where the abbreviation KSTR means the kinematics of the special theory of relativity) (p. 254i):

> It would somewhat clarify McCausland's position if, in his customary lucid style, he would admit in one short sentence that he, Dingle, and Bronowski had failed to notice that

> Einstein's slip, taken literally, immediately contradicts KSTR."

Since some of Good's statements in [8] seemed to be rather questionable, I described in my reply [10] the logical properties of consistent and inconsistent theories, which are very important to the debate. I now present a simpler discussion of these properties.

In order to avoid ambiguity, I shall use the usual symbols of symbolic logic: if we have statements p and q, $p \cdot q$ means p and q, $p \vee q$ means p or q, and $\sim p$ means not p. Suppose that there is a theory **T** having, like special relativity, two postulates, which I call p and q. Now suppose that p and q are inconsistent, that is: they cannot both be true, though they might both be false. I represent that property by saying $\sim(p \cdot q)$. We can represent the properties of this inconsistent theory by writing:

- p
- q
- $\sim(p \cdot q)$

But it follows from the first two postulates that $p \cdot q$. The *inconsistency* has led to a *contradiction*, since $p \cdot q$ and $\sim(p \cdot q)$ contradict each other. Now suppose that we wish to derive some arbitrary proposition s from my theory **T**, we can do it using the disjunctive syllogism in the following logical steps:

$(p \cdot q) \vee s$
$\sim (p \cdot q)$
$\therefore s$

The proposition s can be any statement that we wish, such as "The earth is flat." I have not proved that the earth is flat; I have shown that that statement can be derived, using valid logical rules, from an inconsistent theory. We could equally easily derive the statement "The earth is not flat." Not surprisingly, a contradiction leads to another contradiction. As another example, s could be "Theory **T** is self-consistent."

For example, we could derive from **T** any statement of the form "If you do experiment X you will get result Y." That means that my self-contradictory theory **T** can match any experimental result whatever; for example, it could match all the experimental results that are claimed to support special relativity. This immediately shows that it is not valid to refute a claim of an inconsistency in the special theory by appealing to experimental results, since an inconsistent theory can match the results of all past, present and future experiments.

A contradiction in a theory is an especially pernicious problem. I pointed this out in [10] by supposing that Tweedledum and Tweedledee were debating the question whether a theory **T** is self-consistent. Suppose that Tweedledum derives from **T** two propositions *v* and *w*, which are obviously inconsistent with one another, and claims that the inconsistency is inherent in the theory. Suppose that Tweedledee, in reply, uses a different method to derive ~*v* from T, and claims that Tweedledum is wrong. But, since it is the consistency of **T** that is in question, he has not refuted Tweedledum's claim: *v* and ~*v* could both be derived from an inconsistent theory. In order to refute the claim, he must show that either *v* or *w does not follow* from **T**; that is not the same as showing that either ~*v* or ~*w follows* from **T**, if **T** is self-contradictory. I believe that this was part of the problem in the earlier debate between Dingle and McCrea on whether the special theory entailed asymmetrical ageing. Since both then believed that the theory was self-consistent, each assumed that deriving from the theory a result that contradicted a result that the other had derived was sufficient to refute the other's derivation, whereas they could both have been valid derivations from the inconsistent theory. That is what Dingle meant when he told me that, if the theory were consistent, it would be necessary to disprove one of the two inconsistent results to resolve the paradox, and that showed why he eventually came to believe that the theory was self-contradictory.

As I pointed out in the paper in which I included the above discussion [10], Good was a referee of that paper. My discussion of the properties of inconsistent theories showed him that there was a logical error in his earlier paper [8], and he published an Erratum [9] to correct some statements that he had made in that paper.

Good continued to insist, right up to his paper [11], that E2 was a slip that did not follow from the theory and should not be taken literally. He issued three challenges to me in that paper. The first challenge was to show that E2 did follow from the theory. The second challenge referred to Dingle's claim that synchronization according to Einstein's definition is absolute; Good wrote that "Dingle's assertion was a mistake, and I hope McCausland will now *explicitly* and *clearly* agree." [Italics original.] The third challenge was to state clearly whether Dingle's Question had been answered.

In my reply [13] I showed, by what I think is an ingenious argument, that E2 does follow from E1, and hence it follows from the theory. I also showed that the absoluteness of Einstein's definition of synchronization leads to a contradiction, which I believe helps to explain why most people believe that the definition is not absolute. I also stated that Good had not answered Dingle's question.

Good's next paper [14] bore the title *The Self-Consistency of the Kinematics of Special Relativity, Part VI. Einstein's Argument in Special Relativity Concerning "Time Travel" Is Substantially Correct After All; and Smoothed Polygons*. This is a very strange paper, for various reasons. When he said in his long title that Einstein's argument concerning "Time Travel" is substantially correct after all, he seemed to be referring to the question whether a clock that moves around in a polygonal path or a circular path ends up retarded relative to as clock that remained stationary, which is essentially what is claimed by the statement E1. But that was never in dispute in our debate. The paper G.VI [14] takes about 500 words to define a polygon and to discuss the rounding of the corners to avoid infinite accelerations; these matters are not particularly relevant to our debate, and in any case he had already made the point more briefly when he wrote [8, 252] "Strictly the corners must be somewhat rounded off to avoid infinite accelerations but I will usually ignore that point for the sake of brevity." What more needed to be said?

The first few words of the abstract of [14] read as follows: "What I previously called 'Einstein's Slip' wasn't a slip after all." The abstract went on to say that "This correction doesn't imply an inconsistency in KSTR, although it would appear to do so to those who regard time travel as logically impossible." In the text of the paper he apologized to me for claiming that E2 was a slip on Einstein's part. Since he did not withdraw or repudiate his earlier statement that E2 leads to a contradiction, he was in a highly anomalous position and appears to have conceded that there is a contradiction. I pointed out the inconsistency in his position in another paper [15], which ended our debate since Good died on 5 April 2009.

In his paper [11] Good suggested that, if I did prove that E2 followed from the special theory, my original paper [1] would be resurrected. Now that I have indeed proved that E2 did follow from the theory, and Good has conceded that what he had called "Einstein's Slip" wasn't a slip after all, my original paper still stands unrefuted. Unless someone else refutes my argument, I claim that I have proved an inconsistency in the special theory.

There were other features of the debate that could be discussed, some of which arose from the fact that Good acted as a referee for at least one of my papers, but a full treatment would make this chapter much longer without adding any significant scientific content. I shall therefore refrain from publishing the material unless it later becomes necessary to do so. However, there is one part of my argument that I would like to include here, which I had planned to publish in one of my papers but which, as a result of negative comments made by Good as a referee of my paper [10], I dropped from the published version. I had felt that Good's defence of special relativity relied heavily on the consistency of the mathematical formulation of the theory, which Herbert Dingle had never questioned. The following is a description of the argument that I dropped on Good's recommendation:

I wrote that, although Dingle had made some attempts to give simple illustrations of the possibility that a mathematical result might not have a physical realization in the real world, these were in my opinion somewhat unsatisfactory. I went on to provide what I thought was a better illustration, in terms of a well-known mathematical puzzle. From various published versions of the puzzle, I selected the one by V. Berezinsky (Berezinsky 1973) and rephrased it as follows:

> Three fishermen, after a day's fishing at a remote lake, decide to stay the night there and to divide the catch among them in the morning. However, one of them cannot sleep and decides to go home. He takes his share of the catch by dividing the pile of fish into three equal lots, with one fish left over which he throws into the water, and takes away one of the three lots. In the middle of the night a second fisherman wakes and, not realizing that the first one has already left, takes his share of the remaining fish by dividing the pile into three equal lots, again with one fish left over which he also throws back in the water, and takes away one of the three lots. In the morning the third man, not realizing that the others have left, divides the remaining pile of fish into three equal lots, again with one fish left over. The problem is to find the initial number of fish.

There are many possible answers, the smallest initial number being 25. However, as Berezinsky pointed out, Professor P.A.M. Dirac proposed the following solution: there were −2 fish, and when the first man threw one in the water there were –3, of which he took –1 and left –2 for the next man to continue the process.

I pointed out the obvious fact that, although that solution satisfies the mathematical problem, it does not correspond to any real-world possibility. However, with some imagination and appropriate assumptions it would be possible to rephrase the problem in terms of winnings from a casino instead of fish from a lake, in which case the negative solution might be perfectly feasible; the point is that it is not possible to know, *from the mathematics alone*, whether the solution corresponds to a real-world possibility. Similarly, the consistency of the mathematical structure of the special theory does not in itself guarantee that the physical theory is physically realizable. The following comments on another version of the puzzle (which is more complicated numerically and is described in terms of coconuts instead of fish) has been made by Barrow (1988, 254):

> Here we see an example of a perfectly good *mathematical* solution which (since we cannot realize a 'negative coconut') does not have a realist out-working (however, there is a tradition that Dirac's negative solution played some role in his thinking which led to the introduction of the concept of antimatter). In this situation it is easy to exercise a criterion which eliminates Dirac's solution as an unrealistic one, but in more esoteric areas of mathematical physics it may not be so easy to decide upon criteria by which to reject some mathematical predictions as unrealistic. [Italics in the original.]

Although the above argument was not published during the course of my debate with Good, I still think it needs to be mentioned. Just as negative fish may be consistent with a mathematical equation but do not exist in the real world, so the fact that time travel and multiple universes may be consistent with certain mathematical equations does not necessarily mean that time travel can be accomplished or that multiple universes exist.

References for the Debate in *Physics Essays*

1. I. McCausland: An Inconsistency in Special Relativity, Vol. 3, No. 2, 1990, pp. 176-7.
2. I. J. Good: The Kinematics of Special Relativity, and Dingle's Fallacies, Vol. 4, No. 4, 1991, pp. 591-5.
3. I. McCausland: A Reply to "The Kinematics of Special Relativity, and Dingle's Fallacies," Vol. 6, No. 1, 1993, pp. 127-8.
4. I. J. Good: The Self-Consistency of the Kinematics of Special Relativity, Vol. 7, No. 4, 1994, pp. 436-441.

5. I. McCausland: A Reply to "The Self-Consistency of the Kinematics of Special Relativity," Vol. 9, No. 3, 1996, pp. 484-6.
6. I. J. Good: The Self-Consistency of the Kinematics of Special Relativity, II, Vol. 10, No. 2, 1997, pp. 327-333.
7. I. J. Good: The Self-Consistency of the Kinematics of Special Relativity, III, Vol. 10, No. 3, 1997, pp. 454-465.
8. I. J. Good. The Self-Consistency of the Kinematics of Special Relativity, IV, Vol. 11, No. 2, 1998, pp. 248-263.
9. I. J. Good: Erratum, Vol. 12, No. 1, 1999, p. 190.
10. I. McCausland: On the Consistency or Inconsistency of Special Relativity, Vol. 12, No. 3, 1999, pp. 438-445.
11. I. J. Good: The Self-Consistency of the Kinematics of Special Relativity, Part V(A), Vol. 16, No. 1, 2003, pp. 29-42.
12. I. J. Good: The Self-Consistency of the Kinematics of Special Relativity, Part V(B), Vol. 16, No. 2, 2003, pp. 174-9.
13. I. McCausland: The Persistent Problem of Special Relativity, Vol. 18, No. 4, 2005, pp. 530-532.
14. I. J. Good: The Self-Consistency of the Kinematics of Special Relativity, Part VI. Einstein's Argument in Special Relativity Concerning "Time Travel" Is Substantially Correct After All; and Smoothed Polygons, Vol. 19, No. 3, 2006, pp. 430-433.
15. I. McCausland, On the Inconsistency of Good's Argument Defending Special Relativity, Vol. 22, No. 2, 2009, pp. 81-82.

PART FOUR

Some Unresolved Issues

This part describes some of the confusion that continues to exist in the writings of scientists about relativity. Chapter 9 presents several mutually inconsistent opinions of relativists on the validity of Einstein's statement in his original paper about the relative rates of clocks at the equator and a pole. Chapter 10 examines the relative merits of Lorentz's and Einstein's theories, and questions why one is accepted as representing reality and the other rejected. Chapter 11 points out the important fact that, in answer to a claim that the theory is inconsistent, experimental evidence is irrelevant. Chapter 12 describes some confusion in the literature about the authorship of the footnotes later added to Einstein's paper.

Chapter 9

The Polar and Equatorial Clocks

> Whatever the ultimate decision of historians of later generations might be, as to what Einstein did or did not do, certainly they will agree upon this one thing, namely that he brought more consternation and confusion to the intellectual life of mankind than any human being previously on record.
>
> C. A. Zapffe, (1982, 127).

Einstein's statement about the polar and equatorial clocks has given rise to so much confusion, both in the replies to Dingle's question and in other contexts, that it seems appropriate to devote a whole chapter to the subject.

Let us again look at Dingle's Question. In *Science at the Crossroads*, he quoted Einstein's statement about the polar and equatorial clocks (which I labelled E2 in the previous chapter) with an interpolation as follows (p. 46):

> 'Thence' [*i.e.*, from the theory he had just expounded, which takes no account of possible effects of acceleration, gravitation, or any difference at all between the clocks except their state of uniform motion] 'we conclude that a balance-clock at the equator must go more slowly, by a very small amount, than a precisely similar clock situated at one of the poles under otherwise identical conditions.' Applied to this example, the question is: what entitled Einstein to conclude

> *from his theory* that the equatorial, and not the polar, clock worked more slowly?

Ziman, in his review of *Science at the Crossroads* (1973), omitted the above part of Dingle's statement of his Question, and did not attempt to answer it. As I stated in Chapter 6, Maddox, in his editorial article (Anon., 1972), referred to it by writing

> It seems now to be accepted that Einstein's original argument was uncharacteristically loose. The point of the illustration is that a clock at the pole of rotation may be taken to be in an inertial frame which is nearly (but not quite) properly defined by the direction of the Earth's motion around the Sun. The clock at the equator is in another. Einstein's lack of clarity concerns the inertial frame of the observer of the two clocks.

It is not clear who else has "accepted that Einstein's original argument was uncharacteristically loose." Some other reviewers of Dingle's book accepted that Einstein's statement was valid, and gave their answers based on that assumption. The inertial frame of an observer of the clocks is irrelevant, since it is clear that the effect referred to by Einstein is a real effect and not just something specific to an observer. That is clear from the fact that Einstein added a footnote saying it had to be a balance clock not a pendulum clock; I discuss the relevance of the footnote in Chapter 12.

Stadlen (1972), in his review of *Science at the Crossroads*, also queried the validity of Einstein's statement, and answered the question about the polar and equatorial clocks as follows:

> But the relative motion involved in this case, being circular, is non-uniform. I submit, therefore, that Einstein was wrong in saying that his prediction followed from the special theory, which deals only with the effects of uniform motion. This is not to say that the prediction was invalid. For Einstein was, intuitively, anticipating his later general theory, according to which the equatorial clock runs slower because of the centripetal force exerted upon it.

This is a very strange answer. Stadlen admitted that Einstein's statement was wrong, which conceded Dingle's point that Einstein was *not* entitled to conclude *from the special theory* that the equatorial clock worked more slowly. Stadlen then tried to rescue the prediction by saying that the statement follows from the general theory. That is simply ridiculous; if there is an answer to Dingle's Question, it should have the property that it would have been a valid answer if the question had been asked in 1905, just after the special theory was published and before the general theory appeared. (Perhaps we could extend the date to include the first appearance of the footnote mentioned above (Lorentz *et al.*, 1913); it is still before the general theory appeared.) Even if the slowing does follow from the general theory (which is also a matter of some confusion in the literature), that does not make its prediction from the special theory valid.

It should also be pointed out that anyone who claims that Einstein's statement does not follow from the special theory should be able to show at exactly what point in his argument Einstein made his error. As I showed in *Physics Essays* (McCausland, 2005), the statement E2 follows from E1. Since virtually all relativists accept the truth of E1, they should also accept that E2 follows from E1 or show why it does not follow.

It is interesting to read what is written in another paper about the polar and equatorial clocks, which claims that the general theory *contradicts* the special theory on this subject. Here is an excerpt from the paper by Harvey and Schucking (2005):

> To calculate correctly the rate difference between a clock at a pole and one at the equator, one has to take into account both time dilation and gravitational blueshift. Doing that, one finds that these two effects exactly cancel each other if both clocks are at local sea level.

The paper goes on to refer to the footnote that Einstein added to rule out a pendulum clock, and then goes on to say

> Why this footnote was added is not clear. It is not correct, nor does it address the issue of the correctness of the original prediction.

Harvey and Schucking also appeared surprised that, when Einstein rewrote his 1905 paper in November 1943 for a bond drive, "he even left in the original erroneous prediction, unadorned by any corrective footnote." So these authors again contradicted Stadlen's statement by saying that Einstein's 1905 statement about the polar and equatorial clocks is wrong because the general theory *cancels* it, whereas Stadlen tried to claim that, even if it is wrong, it is valid because the general theory *supports* it.

Another interesting attempt to answer Dingle's question about the equatorial and polar clocks has been made by G. J. Whitrow (1975), in the following statement:

> For a supporter of relativity, the essential difference between the two clocks is that relative to the centre of the Earth (which for the purpose concerned can be regarded as the origin of an inertial frame) the clock at the equator describes a circle and so cannot be associated with an inertial frame, whereas the polar clock is at rest and can be associated with an inertial frame for a period of time during which the curvature of the Earth's orbit can be neglected.

In *The Relativity Question* I answered that statement by writing that there was not a difference in kind between the two clocks, that neither was in an inertial frame, that both moved in approximately circular paths. When Good first wrote to me on the subject (by copy of a letter to Henry Bauer in May 1990), he referred to this argument in the following words:

> McCausland, on page 20, complains that neither A nor B is in an inertial frame. But the acceleration at the pole, due to motion round the Sun is only 0.019 ft/sec/sec, whereas that on the equator is 0.111 ft./sec/sec. So A's frame is much closer to inertial. No real application of theory is exact, so McCausland's complaint is misleading.

I have checked Good's numbers and I agree with them. However, suppose that the speed of rotation of the earth was reduced by a factor of 2.5. Since the acceleration is proportional to the square of the angular velocity, that would reduce the acceleration of the clock at the equator by a factor of 6.25, to about 0.018, while the acceleration of the earth in its orbit was unchanged. The acceleration of the polar clock would then be about the same magnitude as that of the equatorial clock. Furthermore, since the total acceleration of the equatorial clock would be the vector sum of the two component accelerations, there would presumably be a tendency for some partial cancellation effects each day at certain points of the motion, thus making the total acceleration of the equatorial clock small under those conditions; yet Einstein's logic, and Whitrow's, would be unchanged: the equatorial clock would work more slowly. There is nothing in the logic of the special theory to suggest that one compares magnitudes of accelerations to decide which clock works more slowly; if there were, then Dingle's question would be easy to answer.

Another comment on the polar and equatorial clocks was made by Tom Wilkie. This part of the story started with an article entitled "The Twin Paradox Revisited" (Wilkie, 1977) in *Nature*.

Dingle published a short letter (Dingle, 1977b) asking Wilkie to answer a question about his article. Immediately below Professor Dingle's letter there appeared Dr. Wilkie's reply, which did not answer Dingle's question but which said that he was writing to Professor Dingle privately on the matter. As a result of that exchange Dingle wrote a new article on the twin paradox and submitted it to *Nature* for publication, but it was refused. I published the related correspondence (or rather Dingle's side of it, since MacMillan journals would not give me permission to publish their letters) in *The Relativity Question*, and after Dingle's death I arranged for the publication of the article itself (Dingle, 1980).

Although Dr. Wilkie had chosen not to publish a reply to Dingle's letter in *Nature*, he did publish a reply to his posthumous article (Wilkie, 1981). The reply can be read by anyone who is interested, but what seem to me to be two of its most striking features are Dr. Wilkie's deliberate choice *not* to discuss relativity, and his repeated assertions that too much had already been written on the subject of Dingle and his criticism of special relativity. Consider, for example, the following excerpts from his article: "Sir, too much has already been published on the Dingle question, and the time is long past to call a halt to this whole business." "Most academic journals have for some years rightly viewed the matter as settled and regarded more discussion of it as a waste of paper." His final paragraph read: "Too much has been written on this matter already. Please, let it rest." Since he had started the process with his article "The Twin Paradox Revisited" in *Nature*, it seems odd that he should later become so opposed to the revisiting that he had initiated.

Here is what Wilkie wrote in 1981 about Einstein's statement about the equatorial and polar clocks:

> Although I do not have German enough to verify this myself, it may be that there is an error or ambiguity in one of the examples Einstein gave in the paper (comparing clocks at the north pole and the equator). Instead of regarding this, if it be true, as an interesting insight into how Einstein himself had not fully thought out the implications of relativity at that time, Professor Dingle chose to regard that paper as a canonical definition of the theory and used it as the spearhead of his attack. As the discoverer of a possible mistake by Einstein, Professor Dingle might have written an illuminating chapter on the history of science; as Einstein's dogged, but mistaken, critic, he has written himself into that history.

Wilkie did not commit himself to saying that Einstein made an error, but only hinted that he *might* have done so. His statement is completely worthless.

Another point that should be made about the polar and equatorial clocks is that the essential difference between the clocks is that one moves in a circle having its centre at the other; we do not need to picture the situation in terms of a rotating earth orbiting the sun. As Jack Good pointed out in one of his unpublished comments while acting as a referee for my paper in *Physics Essays*, "the earth is a red herring." For example, in a reply to one of Herbert Dingle's arguments, Bronowski (1961) discussed an experiment involving clocks at the centre and the rim of a rotating disc, and related it directly to Einstein's prediction about the equatorial and polar clocks. He argued, in a manner similar to Whitrow's argument discussed above, that the difference in rates of the two clocks is justified by the fact that the equatorial clock is not in an inertial frame. This is how he justified the conclusion that it was the equatorial clock that worked more slowly:

> Relativity only postulates that observers moving in *inertial* systems cannot tell which of them is moving. By contrast, an observer who moves in an *accelerated* system can tell that he has moved, simply by carrying an accelerometer (or a bucket of water).
>
> In the Harwell experiment, the rotating disc is not an inertial system. That is, a point on the circumference is not in uniform motion in a straight line; it is in constant acceleration, which an observer at the point could detect by carrying an accelerometer.
>
> [Italics in the original.] .

Bronowski's argument suffers from the same flaw as Whitrow's: if a clock at the circumference of the rotating disc is in constant acceleration, it is not valid to infer *any* conclusion from the special theory about the rate of that clock. Unfortunately the Editor of *New Scientist* discontinued the correspondence after Bronowski's letter, noting (21 September) that "Professor Dingle wishes it to be known that he does not accept the arguments in Dr. Bronowski's letter."

In his Introduction to Bergson (1965, p. xxvii footnote), Dingle made another comment on the twin paradox, involving twins Peter and Paul, and referred to Bronowski's justification of the asymmetrical ageing as follows:

> Nevertheless during a recent controversy many physicists (for example, J. Bronowski, in *The New Scientist*, Aug. 31, 1961) have continued to maintain that Paul's acceleration on reversal prevents the application of the special theory to the problem. Curiously enough, however, they do not therefore refrain from applying it but regard themselves as entitled to use its equations with a meaning of their own in place of that which the relativity postulate gives them. The result—need it be said?—is that asymmetrical ageing is "proved" to follow from Einstein's special theory. The reader must be left to appraise this procedure for himself.

Bronowski's appeal to the acceleration as the distinguishing feature between the equatorial and polar clocks, with its reference to such simple acceleration detectors as buckets of water, is presumably intended to suggest that anyone who can understand a bucket of water should be able to understand the problem. This approach is similar to that used by Paul Davies (1980) in appealing to acceleration to explain the orthodox resolution of the twin paradox:

> To return to Earth, the twin in the rocket has to turn around, and this involves a period of violent deceleration and acceleration. This is an absolute, not a relative, effect, as testified by the simple observation that eggs on the rocket could be broken during the turn; there is no doubt that it is the rocket, not the Earth, that is accelerating.

However, consider Einstein's original paper (Lorentz *et al.*, 1923), in which he placed two co-ordinate systems with their origins together, each system being "provided with a rigid measuring-rod and a number of clocks," and then suddenly imparted a velocity v to the origin of one of the systems. Obviously, if the systems had been provided with accelerometers or buckets of water in addition to the clocks and measuring rods, these would have clearly shown that one system was accelerated rather than the other, an absolute effect according to Davies; yet Einstein went on to treat the two systems as being equivalent in the sense that it is not possible to say that one of them is moving and not the other.

Although it is not directly related to the scientific world's reactions to Professor Dingle's book, it is interesting to note what another scientist has written on this subject in a review (Anon., 1983) of L. Landau and Yu. Rumer's book *What is the Theory of Relativity*:

> Occasionally, however, in their effort to simplify, the authors make some incorrect statements. The most glaring example

> is in the discussion of the twin paradox, where they have one twin travelling on a large circular railway track, ignoring the fact that the frame of reference is not an inertial one, so is beyond the scope of special relativity.

The twin moving along a circular track has the same status in relation to special relativity as the equatorial clock in Einstein's original paper. If the circular railway track is a glaring error, did Einstein make a glaring error too? As we have seen, there are several contradictory opinions expressed by supposed experts on the subject

I believe that the appeal to acceleration to show a distinction between the two clocks is not valid. In the very first paragraph of his first paper on special relativity (Lorentz *et al.*, 1923), Einstein pointed out that, if a magnet is moved relative to a conductor, or if a conductor is moved relative to a magnet, the current in the conductor is the same in both cases if the relative motion is the same in both cases. Einstein built his whole theory on the fact that *it does not matter* which one moves, the phenomenon that results is the same in both cases; yet, when we are considering two clocks instead of a magnet and a conductor, the theory says that it *does matter* which clock moved. This is a very strange development.

Chapter 10

Einstein or Lorentz?

> After all, the whole of Einstein's theory is wrapped up in the Lorentz transformation, and since the one set of mathematical equations in reciprocal form suffices for both, whatever difference there may be in the underlying physical ideas, it follows that no physical deduction can be obtained from Einstein's theory that cannot also be deduced from that of Lorentz.
>
> E. G. Cullwick (1976).

One of the themes of Herbert Dingle's *Science at the Crossroads* and other critical writings was his comparison of Einstein's theory and Lorentz's. He pointed out that physicists treated the two theories as if they were the same, and used either interpretation at will. So it is important to decide whether in fact either theory can be used in studying physical phenomena.

As I pointed out in my article that appears in Appendix II, Roxburgh (1972) agreed with Dingle to the extent that he said that Lorentz's theory of absolute space, and clock rates dependent on absolute motion, had not been disproved, and he stated that the Lorentz and Einstein theories were "observationally indistinguishable."

Dingle pointed out that Einstein's special theory gained the ascendancy over Lorentz's theory only after the announcement that the results of the 1919 eclipse expeditions supported Einstein's general theory. Another writer who pointed this out was Ohanian (2008, 284) who wrote:

> In Germany, relativity went by the name of "Lorentz-Einstein theory" for some twelve years, and when the name of Lorentz was finally dropped after the end of World War I, this was not because of some recognition or agreement that Einstein deserved the full credit for the invention of special relativity, but because Lorentz was pushed into the background by the stellar fame that Einstein acquired from his theory of general relativity. As Mehra said, Einstein became the great cat of relativity, and everything was attributed to him, rightly or wrongly.

It is interesting to note that Ohanian, even though he wrote a whole book on Einstein's mistakes, made an almost ritualistic disclaimer to dissociate himself from those heretics who dare to criticize relativity. The third sentence of his Preface refers to his book as follows: "And it does not deal at all with the mistakes that misguided souls imagine they perceive in his theories of special and general relativity."

In suggesting that Einstein's special theory overcame Lorentz's because of the announcement of the confirmation of the general theory, it seems to be assumed that the special and general theories are so intimately connected that confirmation of the general should guarantee confirmation of the special; yet John Stachel (1995) wrote as follows:

> 'Theory of relativity' is an umbrella term, covering two quite distinct theories, usually called 'the special' and 'the general' theory. The exact nature of the relation between the two is still a subject of controversy, and both might well not bear the common appellation 'relativity' if not so closely associated with the work of one man, Albert Einstein, who developed the second in attempting to extend the first to include gravitation.

Ohanian also wrote (p. 286) that both the Einstein and the Lorentz-Poincaré ways of dealing with relativity lead to the same results, "but they reach these results by different pathways." He went on to say "Today, physicists prefer to do relativity à la Einstein, because it is more concise and direct, without the extra metaphysical baggage of an undetectable ether and its preferential reference frame absolutely at rest." So it is just a *preference*!

Ohanian went on to write as follows:

> Thus, today, "relativity" is usually taken to mean relativity à la Einstein. However, modern cosmology identifies a preferential rest frame, that is, the rest frame of the cosmic micro-

> wave radiation, or the thermal radiation left over from the Big Bang. And calculations about the cosmic microwave radiation are most conveniently done à la Lorentz, in the preferential reference frame suggested by cosmology. Lorentz would undoubtedly have regarded this as confirmation of his ideas about the ether.
>
> So any dispute over which way of doing relativity is better is ultimately futile. Einstein's is better most of the time, but Lorentz-Poincaré's is better sometimes. And both are right—there is more than one way to skin the cat.

Popper also considered the possibility that special relativity might have to be replaced, and suggested that its place might be taken by Lorentz's theory. Here is what he wrote (Popper, 1982, 30):

> It is only now, in the light of the new experiments stemming from Bell's work, that the suggestion of replacing Einstein's interpretation by Lorentz's can be made. If there is action at a distance, then there is something like absolute space. It should be remembered that nothing like a crucial experiment between Lorentz's theory and special relativity theory has ever been conceived before. If we now have theoretical reasons from quantum theory for introducing absolute simultaneity, then we would have to go back to Lorentz's interpretation.

Popper also wrote (1982, 29) that, even if we had to give up Einstein's interpretation of special relativity and return to Lorentz's interpretation and with it to Newton's absolute space and time, we need not, in that case, give up any formula of special relativity theory.

In the correspondence between Popper and myself, which I described in *The Relativity Question*, this is how he compared the Lorentz and Einstein theories in a letter to me dated 2 September 1980:

> There are so far no crucial experiments envisaged—not even thought experiments—to decide between Einstein's and Lorentz's theories. (My opinion is that Lorentz may be right; should there be action at a distance, then, I claim, we would have to replace Einstein with Lorentz).

Although Popper did not take part publicly in the controversy surrounding Dingle's criticisms of special relativity, he did correspond with Dingle about it, as has been described by Hayes (2010). He was opposed to dogmatism in science, as is shown by a letter

(Popper, 1971) replying to criticism by Dr. G. Burniston Brown, in which he wrote:

> I much prefer any serious critic of Einstein to a man who accepts Einstein's theories because they are, or have been, fashionable, and I have long admired Dr Burniston Brown's intellectual independence in refusing to accept Einstein's theories.

Professor Dirac (1980) has suggested that the postulate of relativity has in fact been disproved by finding an absolute zero of velocity. He contrasted the views of Lorentz and Einstein when he wrote:

> Lorentz had established that so long as one keeps to electromagnetic forces it would be impossible to find an absolute zero of velocity. Einstein went beyond that and said it would never be possible to find an absolute zero in velocity, that there would not be other physical processes that would show up an absolute zero of velocity.

Dirac then went on to make the remarkable statement that Einstein was wrong in saying that it would never be possible to find an absolute zero of velocity; he wrote:

> In one sense, Lorentz was correct and Einstein was wrong, because all Einstein should have said was that with the physics of that time it was impossible for an absolute zero in velocity to show up. But to say that it would never be possible for an absolute zero in velocity to show up was going a bit too far. An absolute zero in velocity has shown up with the more advanced technology which we have at the present day. This refers to the natural microwave radiation.

Although Dirac's statement appears to suggest that Lorentz's theory matches reality better than Einstein's and should therefore be used in its place, he did not reach that conclusion. Here is what he wrote:

> You shouldn't say that Einstein's theory rests solely on its agreement with observation. There is agreement with observation only provided that you don't use a sufficiently advanced technology. In the absolute sense the agreement no longer holds. The real importance of Einstein's work was that he introduced Lorentz transformations as something fundamental in physics. The whole of physics has to be expressed in Minkowski space, a space which is subject to Lorentz transformations. That I would say is the most important of the new ideas introduced by Einstein, and it is of

> tremendous importance and is not disturbed by the more advanced technology which cuts away the basis that Einstein had in proposing his theory.

Why should we believe in Einstein's theory when Einstein was wrong in claiming that it would never be possible to demonstrate an absolute zero of velocity? Dirac gave the following reason:

> I would say the reason why we believe in special relativity is because it puts importance on these Lorentz transformations which are beautiful mathematically. There is certainly no general philosophical basis for it, and we cannot say that it is supported by experiment if we allow experiments involving the most advanced technology.

It seems very strange that we should believe in the *Einstein* theory because of the mathematical beauty of the *Lorentz* transformation, especially when Lorentz was correct and Einstein wrong, and that the real importance of *Einstein's* work was that he introduced *Lorentz* transformations as something fundamental in physics.

We see that both Dirac and Ohanian agreed that the cosmic background radiation provides a preferential rest frame, which favours Lorentz over Einstein. It is therefore interesting to read the different interpretation of the cosmic background radiation by Corey S. Powell (2002, 181). He described how the Nobel Prize was given to Arno Penzias and Robert Wilson "for detecting the pervasive microwave hiss left over from the time of cosmic birth." Powell went on to say that "The notoriously conservative Nobel Committee had now officially converted to the faith of sci/religion," which seems to suggest that the result favoured what Powell called the Church of Einstein. I am sure that the members of the Nobel Committee would be surprised to know that they had converted to that faith.

The similarity of Einstein's and Lorentz's theories is also relevant to one of the few reviews of *Science at the Crossroads* that I did not discuss in *The Relativity Question*. It was published by Johnston (1973) in *The Irish Times*, and it is by far the nastiest review of that book that I have seen. He wrote that the book was an exercise in psychology rather than physics, and wrote that Dingle in his prime was a scientist of standing but that in his decline he had lost touch with reality.

Johnston expressed his admiration for the tolerance exhibited by the "elder statesmen" who had patiently engaged in public controversy with Dingle over the years, "apparently not realising that they were dealing with someone who had become obsessed." He continued with the following reasons for his position:

> My own position is based, not on conservative adulation of elder statesmen, but on having personally been involved in experiments involving measurements of mass and time, which depended so sharply on the reality of relativistic mass-increase and time dilatation effects predicted by the special theory of relativity that they would have been inconceivable without it. If the engineers who built the apparatus had left out from their calculations all "special relativity" terms, the apparatus just would not work.

But it should be obvious from the discussion in this chapter that the apparatus would have worked whether it had been designed using Lorentz's formulation or Einstein's.

I did not discuss Johnston's review in *The Relativity Question* because I did not know of it until I found a copy of it in Dingle's papers in the archives of Imperial College in May 2006. I wonder what his thoughts were as he read the review.

Chapter 11

Experimental Verification of the Special Theory?

> The ultimate argument of the relativists is that the Theory is supposed to have been corroborated by experiments.
>
> Harald Nordenson (1969).

> I hope I shall not shock the experimental physicists too much if I add that it is also a good rule not to put overmuch confidence in the observational results that are put forward *until they have been confirmed by theory*.
>
> A. S. Eddington, (1959).

One of the arguments that Professor Dingle used in various places, to rebut many claims that the special theory had been confirmed by experiment, was that much of the claimed experimental support of the theory depended on circular arguments. There were two main types of argument that he considered to be invalid: these were the interpretation of the Michelson-Morley experiment, and measurements involving the velocities of elementary particles.

For example, Dingle criticized the usual interpretation of the Michelson-Morley experiment (in Bergson, 1965), by saying that it is usually interpreted by saying that the times taken by beams of light to travel different paths were measured, whereas in fact no times were measured and no clocks were used: the times were inferred from electromagnetic theory, and the inference depends on the validity of electromagnetic theory, so the argument is circular.

Dingle also pointed out that, in many experiments that involve elementary particles moving at very high speeds, the speeds of the particles are not measured directly but are inferred from certain observations by a process that involves the use of Maxwell's electromagnetic theory; this fact also rules out experiments of this kind as evidence in support of special relativity.

It is interesting to read some of the very strong claims that the special theory is overwhelmingly supported by experiment. For example, in the book *Was Einstein Right?* (Will, 1986), which deals mainly with the General Theory, there is an Appendix titled "Special Relativity: Beyond a Shadow of a Doubt," in which the author mentions some of the experimental evidence for the theory. One interesting claim is that the evolution of species supports special relativity, because if the time dilation found in special relativity did not occur, muons would not have time to reach the ground and cause the genetic mutations that drive evolution. Whether one is a creationist or an evolutionist, that claim seems to me to be quite a stretch of the imagination.

Another assessment of the experimental support for the special theory was made in the following statement by N. Sperling (2002, 17): "The Special Theory of Relativity is 'only' a 'theory', but it has been tested so many times and so critically that it is something like 99.99999999% likely to be right." In order to justify such an estimate of probability, there would have to have been 10 billion successful experiments performed, with at most one unsuccessful one. It is highly unlikely that so many experiments have been done.

Sperling also estimated the probabilities of some other features of the world (page 18): astrology has a confidence level of 0.00001% (1 part in 10 million) and Santa Claus a confidence level of 0.0001% (1 part in a million). In other words, according to this scientist, astrology is 1,000 times as likely to be right as special relativity is to be wrong, and Santa Claus is 10,000 times as likely.

Although very little notice has been taken of Dingle's claim that the experimental evidence that is taken to support the special theory depends on circular arguments, that may not matter much in the long run, because the circular arguments are not the main problem. The main problem is that, since Dingle claimed that the special theory is self-contradictory, all experimental evidence is irrelevant in answering that claim. As I have shown in Chapter 8, if a theory does have an internal contradiction, one can deduce from that theory, using valid rules of logic, any proposition that one wishes. This property of such a theory has also been pointed out, for example, by Popper (1968, 92): "But the importance of the requirement of consistency will be appreciated if one realizes that a self-contradictory system is uninformative. It is so because any conclusion we please can be derived from it."

That fact has enormous significance in assessing the responses of scientists to Dingle's claim that there is an internal contradiction. A claim that a theory is self-contradictory cannot be refuted by saying that the theory matches various experimental results, since a self-contradictory theory could match all possible results. It does not really matter whether the arguments about the experimental evidence are circular or not; the point is that no experimental evidence whatever can refute a claim of an internal inconsistency.

Sperling's estimate of the reliability of the theory does not impress those who believe that the theory is self-contradictory because, if it was, we would know in advance that the results of any experiment whatever could be derived from the theory. In other words, the reliability of the theory would be 100%.

The question of whether the special theory is self-contradictory can be answered only by studying the theory. Imagine that, in 1905, a suitably-qualified scientist had been confined to a sealed room with a copy of Einstein's paper and asked to find out whether the theory described in the paper was self-consistent. He might not have been able to prove that the theory was self-consistent, but it is quite certain that, if he had found a contradiction, nothing that ever happened outside that room, either the performance of experiments or the publication of the general theory, could remove that contradiction. In short, in assessing the merits of Herbert Dingle's claim that the special theory of relativity is self-contradictory, all experimental results whatever are completely irrelevant.

Many scientists seem to be unaware of this property of logic. Here is an attempt to answer a claim that there is a contradiction in special relativity (Anon., 1967):

> But in circumstances like these, where a theory is lent conviction by the sheer breadth of its agreement with experiment, it would seem incumbent on those who would overthrow it to produce not merely a contradiction but a constructive alternative.

This quote comes from an editorial article in *Nature*, referring to Herbert Dingle's claim that there is a contradiction in the theory. (The article, which was probably written by John Maddox, was reproduced in full as an Appendix in Dingle (1972).) Not only did the writer of this strange argument not appear to understand that the agreement with experiment does not disprove the possible existence of a contradiction, but he also refused to accept that a contradiction—an obviously fatal flaw in a theory—is sufficient to discard the theory unless the person who demonstrates the contradiction also presents an alternative theory.

An argument like this was criticized by Harald Nordenson (1969, 102n), who called it a grotesque argument and made the following comment:

> It is as if a judge, who has established that a person, accused of murder, has brought out evidence of his innocence, would declare: 'It may well be that you have proved your innocence but I cannot release you unless you find who the murderer is.'

The *Nature* editorial about *Science at the Crossroads* (Anon., 1972) was also illogical in that it triumphantly suggested that the Hafele-Keating experiment refuted Dingle's argument, although it did no such thing. One would expect that a scientist who was qualified to write an editorial in *Nature* would know enough elementary logic to know that Dingle's claim cannot be refuted by citing experimental results, but obviously that expectation is not satisfied. It is also relevant to quote a statement made by John Ziman (1984), who wrote as follows:

> Thus, for example, anyone who would now set out to refute relativity theory could not succeed by, say, demonstrating a logical fallacy in Einstein's original paper. It would also be necessary to explain the successful use of this theory in a vast range of scientific work, ranging from the engineering design of particle accelerators to the fundamental equations of quantum physics.

On the contrary, it is obvious that demonstrating a logical fallacy in Einstein's original paper would be fatal for the theory.

An interesting attempt to refute Dingle's arguments by citing experimental evidence was made by Rowan-Robinson (1969a) in a letter in *The Listener* dated August 7. He wrote about what he called "the Herculean task of persuading Professor Dingle that the special Theory of Relativity is self-consistent," and in support of his belief that the theory is self-consistent he went on to list what he called some of the achievements of the theory; these achievements were related to the theory's ability to explain various physical phenomena. But that rebuttal ignores the fact that a theory that was not self-consistent could match all possible physical phenomena, and simply shows that Rowan-Robinson did not understand the logical properties of an inconsistent theory. Then, in a letter in the 4 September issue, W. H. McCrea supported Rowan-Robinson's illogical argument by writing:

> Mr Rowan-Robinson (*The Listener*, 7 August) has given an up-to-date list of some of the main reasons for believing that the physical world does behave in this way.

Rowan-Robinson (1969b) returned to the debate with another letter in the 18 September issue, and again mentioned his earlier letter in the following words:

> Personally I choose special relativity. I do this because it corresponds with experience at so many points, and I listed some of these in my letter of 7 August.

These quotes simply underline Rowan-Robinson's and McCrea's ignorance of the fact that correspondence with experience cannot be used to refute a claim of inconsistency.

In fact, Rowan-Robinson and McCrea, in the correspondence items I have just quoted, are guilty of an even simpler and better-known logical fallacy, known as affirming the consequent. In the words used by Stanley Goldberg (1984, 13) it can be expressed in the following very simple form:

If A then B
I observe B
Therefore A

In Rowan-Robinson's argument, A represents special relativity (assumed to be self-consistent), and B represents the various phenomena that he described and that McCrea agreed with. The form of argument shown in my example is logically fallacious, as pointed out by Goldberg, who wrote:

> It is impossible to show that the explanatory pattern represented by A is unique and the only one to yield the set of results labeled as B.

It seems highly ironic that Professor McCrea, who was Dingle's most frequent opponent throughout the controversies over the special theory, should show such a lack of knowledge of simple logic and of the logical properties of an inconsistent theory.

Another scientist who made an illogical claim that Dingle had been refuted by experiment was Professor Iwan Williams (2003), who wrote about the twin paradox in an article about McCrea:

> Herbert Dingle regarded this paradox as proof of the incorrectness of relativity. McCrea, however, in a famous paper which was published in *Nature*, and is listed in Further Reading, maintained that one twin had accelerated while the other had not, this making the argument invalid, since it is clear that the one that is accelerating is the one that should be regarded as moving. The exchanges between McCrea and Dingle regarding this paradox were many and heated, and have become part of the legend of the early days of relativity. Throughout, though, they remained friendly towards each other, which was a credit to both. (Much later, an experiment was conducted in which an atomic clock was flown around the world. McCrea was duly proved to have been correct.)

The article in *Nature* that is listed by Williams in Further Reading is McCrea (1951). Although the above quotation from Williams refers to the earlier controversy on the clock paradox, even then Dingle and McCrea were debating what the theory required rather than what an experiment would prove, so the experiment did not refute Dingle's arguments about the clock paradox either.

Another writer who mentioned Dingle's criticisms of special relativity was Paul Davies (1995). I first heard about this book from reading a review of it in *Nature* (Landsberg, 1995), in which the reviewer wrote "To my considerable surprise, Davies even resurrects Herbert Dingle's *Science at the Crossroads* (1972), a book whose life expectancy never struck me as more than a few years." I am sure many scientists would like Dingle's book to be consigned to oblivion, but it has not been and will not be forgotten. In his book, Davies quoted Dingle rather scornfully, slightly inaccurately, and in my opinion a little confusingly; the paragraph most relevant to the present essay is on page 57, as follows:

> It has to be admitted that many nonscientists share Dingle's skepticism about drawing profound inferences from a chain of mathematical reasoning, in this case concerning "hypothetical" particles that they can't see and which can be detected only with the help of complicated technology. If time is really dilated, they say, let us observe it on real clocks. As

> luck would have it, just a few months before Dingle's polemical tract appeared, a couple of American physicists managed to do just that.

Davies continued with a discussion of the Hafele-Keating experiment. He did not explicitly state that the experiment refuted Dingle's thesis, but he certainly gave a strong impression that it did. He also discussed "The Puzzle of the Twins," giving the impression that Dingle was still discussing the asymmetrical ageing of the twin paradox. At the end of a description of Dingle's campaign against special relativity in the last years of his life, Davies wrote as follows (p. 55): "The campaign ended abruptly with Dingle's death, but the mood of dissent he championed lives on, widespread and festering. I wonder why? Einstein must have touched a raw nerve."

On the contrary, I think that the reaction of the scientific community to Dingle's criticisms shows rather that Dingle touched a raw nerve. A revealing example of this kind of reaction was given by John Chappell (1980), who described a visit to the physics department of Yale University in the following words:

> The first professor I met listened to me politely for about ten minutes, and then as soon as I had mentioned the name of Herbert Dingle, probably the best-known critic of Einstein then living, he lost his composure and flew into a rage, and loudly demanded that I get out of his office at once.

Chappell described the meeting equally graphically in a letter to Dingle dated 8 August 1969, in which he tried to persuade Dingle to submit a paper to a conference he was organizing. He wrote:

> The name of Herbert Dingle is one I would much like to feature. No other can evoke a purple rage in a calm physicist and cause him to order you out of his room after theretofore being extremely, but haughtily, polite. (Prof. Lamb of Yale did that when I mentioned your name.)

If it seems surprising that a professor at an American university would become so emotional about Herbert Dingle, it may be noted that Willis Lamb was a professor at Oxford University for some years.

Another suggestion that Dingle may have touched a raw nerve appeared in an obituary notice by S. Jocelyn Burnell (1978), in which she told how her physics lecturer said "look at the work of R. B. Dingle. That's not the Dingle who misinterpreted the theory of relativity." That was the first time that she had heard of Herbert

Dingle, and she was struck by the "uncharacteristic vehemence" of the lecturer's remark. Burnell's obituary notice ended with the following paragraph:

> He was a man of considerable intellect with an excellent command of the English language (including the capacity for sharp retort). He leaves his lists unvictorious but unvanquished. In years to come, who knows? He might be right yet.

Returning to the question of experimental evidence and self-contradictory theories, it should be pointed out that another problem with a self-contradictory theory is that it is immune from being refuted by experiment because, whatever the outcome of an experiment might be, it would be possible to derive a result from the theory that would match the result of the experiment. I think we should heed the words of Popper (1982, 198), who wrote that irrefutability is not a virtue but a vice.

However, although the special theory would be immune to refutation by experiment if it were self-contradictory, it might still be possible to disprove by experiment one or other of the two postulates of the theory. In fact, as I pointed out in the previous chapter, Dirac suggested that the first postulate, the postulate of relativity, had been disproved. It might also be possible to disprove Einstein's second postulate by experiment, and Dingle suggested such an experiment (Dingle, 1972, 210).

After the foregoing part of this chapter was written, it was announced in September 2011 that experiments by the OPERA (Oscillation Project with Emulsion-tRacking Apparatus) collaboration had detected neutrinos travelling faster than light through a 730-km tunnel from CERN in Switzerland to Gran Sasso in Italy. In these experiments the speeds were measured using clock readings at the two ends of the path, not just by inference from electromagnetic theory. The experiment appeared to disprove special relativity, according to which no physical object can travel faster than light.

The results of these experiments have no more relevance to Dingle's thesis than the results of experiments that support the theory, since they have no bearing on whether there is or is not a contradiction in the theory. However, discussion by scientists and others after the results of the experiments were made public have added a very interesting topic to the story. Many scientists expressed extreme scepticism about the results. For example, according to a report in *Nature* (Reich, 2011), "Physicists, for the most part, suspect that an unknown systematic error lies behind OPERA's startling result." A very striking reaction to the news appeared in an article in *The Times* of London (Al-Khalili, 2011) in which a Professor of Physics at the University of Surrey wrote "If the experiment proves to be correct and neutrinos have broken the speed of light, I will eat my boxer shorts on live TV." The writer tried to back both sides of the question by writing as the last paragraph of his article:

> Even though scientists will remain sceptical for now, they, along with everyone else, would most love for this revolutionary result to be true. Even Einstein would love it.

Some other scientists reacted to the news of the experiments by suggesting that the neutrinos might have taken a short cut through one of the extra dimensions of a higher-dimensional space. If that kind of argument is permitted, it would indeed be difficult to use the results of the experiment to disprove the special theory.

I find it interesting to compare the reactions of scientists in 2011 to the news of the faster-than-light neutrinos to the reactions of scientists in 1925 to the claim that D. C. Miller had obtained experimental results contrary to the Michelson-Morley experiment. Polanyi (1958) has described how at that time a layman might have expected scientists at least to suspend judgment until the results might be accounted for without disproving relativity, and commented as follows:

> But no: by that time they had so well closed their minds to any suggestion which threatened the new rationality achieved by Einstein's world-picture, that it was almost impossible for them to think again in different terms. Little attention was paid to the experiments, the evidence being set aside in the hope that it would one day turn out to be wrong.

Without expressing any opinion on the reliability of either of the two sets of experiments, I suggest that my quotation from Polanyi would be a fair description of the early reactions of scientists to the news of the faster-than-light neutrinos.

On the question of whether Dingle's claim of a contradiction in special relativity can be refuted by experimental evidence, I believe that the situation can be described very simply as follows: the presence or absence of a contradiction in special relativity is a property of an organized collection of words and symbols that were printed on certain pages of a certain journal in 1905; that property cannot be decided by a physics experiment, any more than the presence or absence of a spelling error in the same paper could be decided by such an experiment.

Chapter 12

Footnotes to Einstein's Paper

> Routine footnotes usually provide welcome information, but more often than not the principal pleasure they give is to distract the reader from the text he is reading. If the text should actually prove to be absorbing, ordinary footnotes afford no pleasure whatever. Encountering one, as Noel Coward remarked, is like going downstairs to answer the doorbell while making love.
>
> G. W. Bowersock (1984)

When Herbert Dingle was discussing Einstein's statement about the polar and equatorial clocks, which he had included in the statement of his Question which I quoted in Chapter 5, he added the following comment (Dingle, 1972, 237):

> Moreover, he added a footnote to say that the result did not apply to a pendulum clock. This would have been meaningless if it were not the actual physical working of the clocks that was in question but merely an accident of the observer's standpoint.

An important fact about that footnote is that it did not appear in the original version of Einstein's paper on special relativity as it was published in *Annalen der Physik*, but was one of several footnotes added later when the paper was included in a collection of papers on relativity. The fact that the footnote was added later appears to make the statement even more emphatic. However, for some strange reason, several people have stated that the added footnotes were written by Arnold Sommerfeld (1868-1951). Since there seemed to be no reason whatever for Sommerfeld to have written the added footnotes while Einstein was still available to write his own notes, I eventually decided to look for evidence for the attribution. As a result of my investigation of the subject, I came to believe that there is no evidence whatever that Sommerfeld wrote the added footnotes, and I published a short article on the subject (McCausland, 1984). Since one of the footnotes had been attributed to Sommerfeld in a paper by Cullwick (1981) in the *British Journal for the Philosophy of Science*, I published my argument as a discussion item in that Journal. I pointed out that the footnotes in question had first appeared in German in a collection of papers that was published in 1913 (Lorentz *et al.*, 1913), and that the editor of that collection, Otto Blumenthal, had acknowledged the friendly co-operation of Einstein in preparing the collection; I also gave various other reasons for attributing the footnotes to Einstein himself.

To my surprise and amusement, as a result of my brief essay I have become a footnote to Einstein myself, for my article was cited in *The Collected Papers of Albert Einstein* (Einstein, 1989, 307). After making specific reference to one of the footnotes, the note referring to my article ends as follows:

> The authorship of these unsigned additional notes is unclear (*see McCausland 1984*). The title page of *Blumenthal 1913* says "A Collection of Papers with Annotations by A. Sommerfeld" ("Eine Sammlung von Abhandlungen mit Anmerkungen von A. Sommerfeld"). However, the extensive notes by Sommerfeld on Minkowski 1909, printed after the reprint of that paper, are listed separately, under Sommerfeld's name, in the table of contents. Lorentz signed a note he added to the German translation of Lorentz 1904a. If Einstein did not write the additional notes to this paper, the contents of some suggest that he was consulted.

The above note was unsigned, but was probably written by John Stachel, who was Editor of that volume of *The Collected Papers*. If he did not write it, I think it is reasonable to assume that he was consulted. Unfortunately the note confuses the issue in several ways. The expression "A Collection of Papers with Annotations by A. Sommerfeld" gives a misleading impression. Although the German words are quoted in their correct order, the significance of the layout of those words on the title page has been ignored. On the title page of *Blumenthal 1913* (my Lorentz *et al.*, 1913) the words "*Eine*

Sammlung von Abhandlungen" appear as one line, and the words "*Mit Anmerkungen von A. Sommerfeld / und Vorwort von O. Blumenthal*" appear as a two-line expression below; both expressions are in capitals and the two-line expression is in a smaller typeface than the first. That shows that the two expressions refer to two different things: the collection of papers, and the contribution by Sommerfeld.

As is suggested in the Note in the Einstein volume, the Table of Contents of the 1913 collection clearly associates Sommerfeld's name with the notes on Minkowski's paper. The promise made on the Title Page and in the Table of Contents, that the collection includes *Anmerkungen* by Sommerfeld, is completely fulfilled by the appearance of the five-page article headed "*Anmerkungen*"; there is no need to look for any further contribution by Sommerfeld.

I suggest that it is completely inappropriate for Sommerfeld's five-page article to be described as "Annotations." That word usually refers to brief separate notes, rather than a five-page article such a Sommerfeld's. I think that a person who was involved in the editing of Einstein's papers, writing a comment on my essay, should have known that "*Anmerkungen*" was translated as "Notes" on the title page, in the table of contents, and in the heading of Sommerfeld's article, when the 1922 edition of the collection of papers was translated into English (Lorentz *et al.*, 1923).

As I pointed out in my original note about the footnotes, the probable reason for Sommerfeld's writing notes about Minkowski's paper and not the others was that Minkowski was the only author of a paper in the 1913 collection who was no longer living at that time. Minkowski was also treated differently in having his photograph appear as the Frontispiece of the 1913 collection, this special treatment being described in Blumenthal's Foreword to that collection as "*das Vorrecht der Toten*" (the prerogative of the dead).

Presumably the comment about Lorentz signing his added footnote is meant to suggest that, because Einstein did not "sign" his added footnote, there is some doubt that he wrote it. But the comparison of Einstein's paper and Lorentz's is completely irrelevant, because of the different histories of their papers. The citation for Lorentz's paper (Einstein, 1989, 609) shows that it was originally published in Dutch and then in English, both in Amsterdam, in 1904. Then, when it was published in German in the first collection of papers in 1913, Lorentz added the footnote mentioned above. Lorentz's footnote was a substantial addition to his paper, about 140 words, and so it was reasonable for him to add his name to that note, which is much longer than any of the other footnotes I am discussing. When the English translation of the collection of papers subsequently appeared (Lorentz *et al.*, 1923), the earlier English 1904 version of Lorentz's paper was included, *without his added footnote*, in the collection. So Lorentz's paper is somewhat anomalous, and the mention of his having signed his footnote has no relevance whatever to the discussion of the footnotes to Einstein's paper. In any case, if Einstein did not sign the footnotes in question, neither did Sommerfeld, so the evidence provided by the reference to Lorentz is precisely nil. Furthermore, since four of the footnotes in the republished version of the paper were from the original 1905 version of the paper and five were not, I suggest that, if the new footnotes had in fact been written by someone other than Einstein, the Editor of the collection of papers would have had a responsibility to identify their author explicitly.

In his book *Einstein's Miraculous Year*, (Stachel, 1998, 159-60) John Stachel wrote in an Editorial Note on the subject of the footnotes, "If Einstein did not write the additional notes to this paper, the contents of some of the notes suggest that he was consulted." In the absence of the commentary that appeared in the *Collected Papers*, (Einstein, 1989) and without any indication that there is any doubt about the matter, readers of the book may find that statement rather puzzling. The idea that someone might have added some footnotes to Einstein's paper without consulting him seems too absurd to require rebuttal.

The authorship of the footnotes is in itself relatively unimportant; a problem arises if someone uses belief in another authorship either to weaken or to strengthen the message in the footnote to support his own predilections. See, for example, Holton's paper (Holton, 1960) in which he wrote as follows, referring to something that Whittaker had written:

> Whittaker clearly implied that Einstein used Lorentz's transformation equation published in 1904. He therefore chose to neglect that *both Einstein and those close to him* have repeatedly said that Einstein had not read Lorentz's 1904 paper. [Italics mine.]

In support of that statement he cited "the footnote on this point by A. Sommerfeld." This refers to a footnote that reads "The preceding memoir by Lorentz was not at this time known to the author." As I had pointed out in my original published note, the information contained in that footnote could have been provided only by Einstein himself. Holton assumed that the footnote was by Sommerfeld and wrote as if Sommerfeld had provided independent corroborative evidence supporting Holton's point of view, even though the only reliable source of information on that subject was Einstein himself. It is interesting to note that Holton also referred elsewhere (Holton, 1993) to yet another of the added footnotes by saying that it was "added by the editor when the paper was reprinted." Otto Blumenthal was the editor, not Sommerfeld, so this confuses the issue even more. Would Blumenthal also be taken to have corroborated the information provided in the footnote?

On the other hand, if someone wished to disagree with one of the footnotes he could try to dismiss it by saying that Einstein had not written it. For example, as I have pointed out in Chapter 8, Good argued that Einstein's statement about the polar and equatorial clocks was a slip on Einstein's part. Good's description of Einstein's statement, on page 252 of one of his papers (Reference [8] of Chapter 8), cites the source of the statement (his Reference 25) in such a way as to include "a note by A. Sommerfeld." This seems to imply that Good believed that Sommerfeld added the footnote mentioned above, even though I had already mentioned my note on the subject in my first paper of the debate (Appendix IV). In another paper, otherwise unrelated to our debate, Good (1998) wrote that, according to a book by Arthur I. Miller, another of the footnotes was by Sommerfeld.

It is difficult to know how the belief arose that Sommerfeld added the footnotes that have been attributed to him. The earliest suggestion I have seen that Sommerfeld may have written them is in a book by Whitrow (1967). On page 93 we read:

> Einstein's most important original papers on relativity, edited with notes by A. Sommerfeld, are available in English translation in
>
> A. Einstein and others, *The Principle of Relativity*, translated by W. Perrett and G. B. Jeffery, London, Methuen, 1923.

The wording, and the layout of the two paragraphs, make it clear that Whitrow's expression "edited with notes by A. Sommerfeld" refers to Einstein's papers, even though there is no evidence to suggest that Sommerfeld wrote notes on Einstein's papers.

A paper by Max Jammer (Jammer 2004), refers to my published note about the footnotes. He says, referring to one footnote that appeared when the 1905 paper was reprinted in 1913:

> It has been usually assumed that Arnold Sommerfeld, on whose initiative this Collection was printed, wrote this footnote. But as McCausland has recently shown, it was most probably Einstein himself who wrote it or was at least consulted whether it should be added.

Another paper that mentions the footnote about the clocks is by Harvey and Schucking (2005). As I mentioned in Chapter 9, the paper claims that the footnote is wrong, and indeed that the argument is wrong because it is contradicted by the general theory. However, the paper does say that I made a convincing case about the authorship of the footnotes.

PART FIVE

Problems in the Scientific Community

This part describes why I believe it is important for the scientific community to take note of the problems raised by the confusing and dogmatic attitude of members of that community in their attitude to relativity. Chapter 13 describes how a very eminent professor of physics used a book review to denigrate those who criticize relativity, but when he was challenged it became clear that he was not an expert in the subject himself but was relying on his belief in the soundness of the views of other scientists. Chapter 14 argues that the scientific community tends to resist criticism of what are generally believed to be well-established theories. Chapter 15 describes some of my own experience in having critical papers rejected for reasons that were obviously unrelated to the merits of the papers in question. The epilogue summarizes my personal opinions on the present situation.

Chapter 13

A Cavendish Professor

> It is not what the man of science believes that distinguishes him, but how and why he believes it. His beliefs are tentative, not dogmatic; they are based on evidence, not on authority or intuition.
>
> Bertrand Russell, *A History of Western Philosophy*

In 1983 there appeared a review of Abraham Pais's biography of Einstein, "*Subtle is the Lord...*" (Pippard, 1983). Professor Pippard (1920-2008) was identified by the statement "Sir Brian Pippard was until recently Cavendish Professor of Physics at the University of Cambridge." For some reason he chose to include in his review an attack on anti-relativists, part of which ran as follows:

> The dogmatic certainty of physicists concerning Special Relativity is a source of sorrow to the less sophisticated, who are unable to swallow the "Twin Paradox" (not Einstein's term). Even if there were no nuclear experiments to verify that it is possible to embark on a long space voyage and return less aged than the stay-at-home, the relativist would still have to reject his critics, whose principal error is to suppose that common sense still rules in the world of relativity. ... From now on he must be careful to apply mathematical logic rather than common-sense intuition to the solution of problems, until he has redeveloped his physical intuition to include relativity.
>
> This is the step the objectors cannot take, but for those who can the rewards are immense. Henceforth every theory must

> be compatible with relativity, and on important occasions this has enabled rival views to be critically assessed.

It should not be necessary to point out that, even if asymmetrical ageing is shown by nuclear experiments, it is not valid to extrapolate from those experiments to say that it is possible for humans to go on long space voyages and experience significant slowing of their ageing. It should also be noted that the arguments of the critics of the twin paradox, like Herbert Dingle, were not about whether asymmetrical ageing actually occurs but whether it is a requirement of the special theory.

Pippard went on to say that "it is hardly surprising that the physicist adopts relativity theory as one of his certainties, knowing that only chaos, not enlightenment, awaits any attempt to discard it."

After reading Pippard's review I felt that the attack on anti-relativists was completely unjustified, and wrote the following letter to the Editor, who published it in the April 22 issue (McCausland, 1983a):

> It is a little difficult to know why Sir Brian Pippard devotes such a large proportion of his review of Abraham Pais's book "*Subtle is the Lord* ..." (April 1) to an attack on anti-relativists. He does not refute their case, but sets up his own interpretations of some anti-relativist arguments in order to dismiss them, and also accuses the critics of incompetence ("This is the step the objectors cannot take ...").
>
> It is clear, both from the wording of Sir Brian's argument and its context in the review, that he assumes that the twin paradox (or clock paradox) and the asymmetrical ageing of space travellers can be adequately treated within the domain of special relativity. Yet the book that he is reviewing includes the following statement (p. 145), referring to Einstein's original prediction of asymmetrical ageing in his paper on special relativity:
>
>> He [Einstein] called this result a theorem and cannot be held responsible for the misnomer *clock paradox*, which is of later vintage. However, as Einstein himself explained some time later, the logic of special relativity does not suffice for the explanation of this phenomenon (which has since so often been observed in the laboratory) since frames other than inertial ones come into play.
>
> If the result in question is a theorem of special relativity, and yet the special theory is insufficient for the explanation of

> the phenomenon, then these facts themselves show that the special theory is inadequate.
>
> Sir Brian may be correct in saying that any attempt to discard the special theory would cause chaos rather than enlightenment. That is not sufficient reason to refuse to consider that possibility, and I suggest that scientists should take to heart Emerson's statement that we can take our choice between truth and repose, but that we can never have both. The abandoning of special relativity would involve a scientific revolution; like other scientific revolutions, it might cause chaos for a time, but it might also lead to an enormously stimulating period of scientific research. Scientists should not shrink from grasping such an opportunity.

I hoped for a reply from Professor Pippard, and I was disappointed that no reply to my letter was published. When it became clear that Pippard would not reply, I addressed the following letter, dated May 19, 1983, to the Editor:

> Sir Brian Pippard's reply to my earlier letter (Einstein, April 22) uses the strongest and most frequently used argument of the relativists, an argument which is singularly difficult to rebut: complete silence.
>
> There are two reasons why silence is an effective argument in this case. Firstly, if an obscure engineer criticizes a former Cavendish Professor's views on relativity, without reply, it will be widely assumed by readers that the critic is a crank whose views may be safely ignored. Secondly, the published arguments that have been used to defend special relativity against its critics are so diverse and mutually inconsistent (as I have documented elsewhere) that it is almost impossible for a scientist to make any statement in defence of the theory without contradicting what some other defender has written; in such circumstances silence is the only effective response.
>
> My earlier letter mentioned a contradiction; please permit me, Sir, to state it more explicitly. With reference to the phenomenon of asymmetrical ageing, generally associated with the terms twin paradox and clock paradox, consider the following pair of propositions:.
>
> 1. Einstein's special theory is sufficient for the explanation of the phenomenon.
> 2. Einstein's special theory is not sufficient for the explanation of the phenomenon.
>
> The first proposition represents the position taken by modern textbooks on special relativity and, as I suggested in my

> earlier letter, apparently supported by Sir Brian Pippard. The second proposition, according to the passage from Pais quoted in my earlier letter, represents the position taken by Einstein.
>
> Since the two propositions are logically contradictory, one of them must be false. I invite Sir Brian Pippard to state, clearly and explicitly, which one is false. I look forward with interest to his reply.

The Editor declined to publish this letter; that is not surprising, since it is not usual to publish one side only of a debate in such correspondence columns. However, another scientist, Professor Richard A. Waldron, a former student of Professor Pippard, wrote to him to protest against his attitude to those who could not accept the special theory of relativity. Professor Waldron (1925-1990) was Head of the School of Mathematics at the Northern Ireland Polytechnic (now part of the University of Ulster), and the author of a book *The Wave and Ballistic Theories of Light: A Critical Review* (Waldron, 1977) which is mentioned in the correspondence between himself and Pippard. My correspondence with Professor Waldron had started when he wrote to me to express solidarity after seeing Herbert Dingle's posthumous article (Dingle, 1980) that I had arranged to have published. Since he had also sent me a copy of some newspaper correspondence in which he had participated, I wrote to him to draw attention to Pippard's review. He decided to write to Pippard rather than to the TLS, and sent me a copy of the correspondence. Part of his first letter to Pippard, dated 3rd May 1983, is as follows:

> I don't understand how you can be so intolerant of those who cannot accept STR. Would you include among the "less sophisticated" such people as Ives, Palacios, O'Rahilly, Cullwick, Burniston Brown, Essen, Dingle? The reason they could not "swallow" the "twins paradox" is that it is not a paradox but a contradiction, and a contradiction in a theory makes it untenable. This fact cannot be overcome by "dogmatic certainty," for dogma has no place in science; indeed, for anyone but a theologian to be dogmatic is tantamount to an admission of error. If STR is a matter of dogma, then there is nothing to discuss, but if it is proposed as a scientific theory, its upholders must answer its critics, not ridicule them. ("When you haven't got a case, abuse the opposing counsel").

Professor Pippard replied to Professor Waldron on 10th May; the first paragraph of his reply was as follows:

> Of course I am ready to consider a variety of opinions, but that does not mean that I must take seriously a point of view which I hold to be wrong. You ask me whether I would consider among the "less sophisticated" a number of people you name. Well, yes, I would. When I compare them with Einstein, Pauli, Bondi, Hawking, Hewish, Møller, Heisenberg *etc. etc.*, I regard them as country bumpkins. I am not an expert at relativity myself, but I know the intellectual quality of those who have thought deeply on the matter, and have not the slightest doubt that if there were a flaw so obvious that Essen or Dingle could find it, they would have exposed it long ago.

The letter went on to discuss some theoretical points, which are not relevant to my story. I find it interesting that none of the scientists mentioned in Pippard's letter had attempted to refute Dingle's thesis. Professor Waldron replied by letter dated 16th May; after some technical comments that are not important to my story, the main point of his letter was worded as follows:

> The essential point is that many people have pointed out contradictions in STR. If the theory is to stand, its upholders must show how those contradictions can be resolved. Instead of our arguments being refuted, however, we are ignored (you have ignored the point on pp. 73-77 of my book that I drew attention to in my last letter), or ridiculed ("less sophisticated," "country bumpkins," *etc.*), or persecuted. The extent of the problem is exemplified by Dingle's *Science at the Crossroads*, which gives his own story. My own is fairly similar, and I know of a large number of other cases, not only in this country.
>
> Is such behaviour in the best tradition of scientific practice? Is science guided in the direction of healthy progress by the ruthless suppression of unorthodox opinion, or is it shunted into a *cul-de-sac*?

Professor Pippard replied to Professor Waldron in a letter dated 29 May; the first part of his letter is as follows:

> I don't think we shall resolve our differences by correspondence, especially as I have never studied relativity, nor taught it. I remember well the *Nature* correspondence of 1956 or 1957 that you refer to in your book, and the exasperation of my learned friends as they deployed all the ways they could think of to convince Dingle that there was no logical incon-

> sistency in STR. When a conflict like this arises, only time will resolve it. There is no question of ruthless suppression; it is rather that the upholders of STR are persuaded that the paradoxes are inevitable, just as the paradoxes of quantum mechanics have resisted all attempts at resolution. We teach these things because we have enormous bodies of evidence that they provide the techniques to get the right answer, and that there is no systematic alternative which does the same in a philosophically more acceptable way. I'm afraid your ballistic theory of light has not got the internal coherence to provide a substitute.

Professor Pippard went on to discuss some specific problems in Waldron's book about the ballistic theory of light, the details of which are not relevant to my story. One minor remark that might be made about his letter is that Professor Pippard's memory seems to be somewhat inaccurate in the dating of the controversy: Dingle was not claiming in 1956 or 1957 that there was a logical inconsistency in the theory, but simply attempting to convince scientists that the orthodox resolution of the twin paradox was incorrect; he did not make the claim of logical inconsistency until much later.

It is unfortunate that Pippard's prestige as a former Cavendish Professor was used to authenticate the statements in his review article, when he was not an expert on the subject he was writing about in the part of his review that we are discussing. His argument for believing relativity is like saying Einstein must be right because all those eminent scientists have endorsed him, which is a totally unscientific argument.

I would like to pay tribute here to Professor Pippard's willingness to allow me to quote from his letters, a willingness that I have found to be rare among those who have criticized the critics of relativity. When I wrote to him in March 1994 to ask permission to quote from his letters to Professor Waldron, he immediately responded in a very positive and friendly way, saying "I find nothing in my letters that I wish to retract (certainly not my reiterated disclaimer of any but superficial knowledge of relativity) and I have no objection to being quoted if that is what you wish." He did have reservations about being quoted about part of the technical argument in one of his letters, but that does not affect my discussion.

Since Pippard's reply was so positive and open, when I replied I took the liberty of sending him copies of my two letters to the TLS, the one that was published and the one that was not, and asking for his comments. Unfortunately he relapsed into silence again, and I did not receive any reply to that communication.

Professor Pippard was, of course, free to use whatever criteria he wished to form his own private opinions. It was when he publicly denigrated the critics of relativity, using the prestige of his Cambridge professorship but relying only on the views of other scientists, that his actions became unscientific and reprehensible.

Chapter 14

Science and Criticism

> That dissent should be regarded as disloyalty, criticism as an unfriendly act, difference of opinion as disaffection, show how far we have drifted in the direction of authoritarianism and away from the democratic principles which we profess—principles according to which free and frank and critical discussion of all questions should be not only tolerated but encouraged.
>
> A. Macbeath (1952).

On 6 June 1947 Herbert Dingle delivered his Inaugural Lecture, on *The Missing Factor in Science*, as Professor of History and Philosophy of Science at University College, London. He said (Dingle, 1952, 8) that "the thesis I wish to propose is that the state of unselfconscious automatism in which science finds itself today is due to the lack throughout its history of a critical school, working within the scientific movement itself and performing the function, or at least one of the functions, which criticism has performed for literature from the earliest times." He emphasized the need for a critical effort within the scientific movement, and pointed out that most of the critics of science have not been within the scientific movement. In fact, he wrote "Most of them have been opponents of it, and their aim has been in the main not so much to understand as to discredit, not to appreciate the significance but to expose the defects of the movement."

Dingle's concerns were echoed by Stanley Jaki (1924-2009), who described Dingle's lecture as "a penetrating analysis of this baffling situation" and went on to say (Jaki, 1988):

> Due to the absence of a school of criticism within science, the errors, illusions, half-truths, and wishful thinking of science are not aired vigorously within the scientific community. Consequently, many in the scientific community come under the illusion that the possession of truth is their special privilege.

Jaki wrote of scientists that "They may deplore the shortsightedness of bygone generations of scientists, yet they are curiously insensitive to their own." A similar point was made by Dingle in a letter to Michael Rowan-Robinson in September 1969:

> One of the most serious results of the neglect of the history of science is the failure to see that its progress continues from its beginning to a point indefinitely far into the future, and we happen to be at one point, in no way fundamentally distinguishable from any other, at which we hold certain ideas. To suppose that they are intrinsically any more permanent than those held at any other point in time is just to be parochial.

Rowan-Robinson himself showed his awareness of the impermanence of scientific knowledge when he wrote as follows (Rowan-Robinson, 1996): "I believe it is important to emphasize that science is to do with argument and controversy and is not a sacred canon of 'knowledge' handed on from generation to generation."

Dingle's own criticisms of science were many and varied, over a very long career. His criticisms arose from his love of science; that is why he was concerned that so many of the critics of science were opponents of science, which led to the tendency of scientists to resent criticism so that they reject it or ignore it.

An interesting example of a scientist's attitude to criticism is shown by the following statement by G. H. Hardy (1877-1947) in his well-known book *A Mathematician's Apology* (Hardy, 1940):

> Statesmen despise publicists, painters despise art-critics, and physiologists, physicists, or mathematicians have usually similar feelings; there is no scorn more profound, or on the whole more justifiable, than that of the men who make for the men who explain. Exposition, criticism, appreciation, is work for second-rate minds.

Even if criticism is work for second-rate minds, it is a necessary service that ought not to be scorned. Even Hardy himself wrote at least one book review, of Whitehead and Russell's *Principia Mathematica*, in the *Times Literary Supplement*. Although the review was anonymous, as was the custom in those days, he was identified as the reviewer by D. May (2001).

As Oscar Wilde wrote, in his essay *The Critic as Artist*, criticism can be a highly creative activity, perhaps even more creative than the discovery of new knowledge. I believe that this can be true in science. One might ask, for example, whether the creative contribution of a scientist chosen at random from the list of more than 250 co-authors of a single scientific paper by F. Abe *et al.* (1991) was greater than, for example, the creative contribution of Herbert Dingle's Inaugural Lecture or his Presidential Address to the Royal Astronomical Society.

T. H. Huxley (1894) wrote that the essence of science is criticism. However, in modern times many scientists are extremely resistant to criticism. Another scientist who expressed a dislike of criticism was Gerald Holton, who wrote the following as the first sentence of Chapter 6 (The Anti-Science Phenomenon) of his book *Science and Anti-Science* (Holton, 1993):

> Opposition to science as conventionally defined can take a great variety of forms, from interest in astrology to attacks on relativity theory, from false beliefs based on scientific illiteracy to support of Lysenkoism or Creationism.

Why should criticisms of relativity theory be described as attacks on relativity theory? Are publications like the *Times Literary Supplement* and the *New York Review of Books* filled with attacks on the books that are reviewed? Holton was aware of Dingle's criticisms of special relativity: in a review (Holton, 1992) of a book by Klaus Hentschel he wrote that "Hentschel seems a little too gentle with some of the theoretical physicists (*e.g.*, H. Dingle and E. Whittaker) who managed to persist in legendary wrongheadedness towards Einstein's SRT." Whittaker's sin was his famous statement (Whittaker, 1953) that "Einstein published a paper which set forth the relativity theory of Poincaré and Lorentz with some amplifications, and which attracted much attention." That statement appeared in a chapter bearing the title "The Relativity Theory of Poincaré and Lorentz." Needless to say, it is not viewed favourably by admirers of Einstein's relativity.

Not only is criticism a creative activity, it can be helpful to the people being criticized if they are able to accept it, as was vividly described by Randy Pausch in his famous Last Lecture; this is how it was worded on page 37 of the book (Pausch, 2008) which was high on best-seller lists for well over a year:

> When you see yourself doing something badly and nobody's bothering to tell you anymore, that's a bad place to be. You may not want to hear it, but your critics are often the ones telling you they still love you and care about you, and want to make you better.

Would those who, like Holton, consider criticism of relativity to be anti-science really classify as anti-science the critical writings of Herbert Dingle, whose love of the scientific adventure was obvious in almost everything he wrote? It would be an insult to him to classify his criticisms of special relativity with interest in astrology and support of Lysenkoism or Creationism and to call them all anti-science.

I will now give an example to show the scientific community's intolerance of criticism of relativity in particular. As I have mentioned before, Dr. Guy Burniston Brown (1902-1988) was a severe critic of relativity for many years, long before Herbert Dingle became convinced of the untenability of the theory, and he published in the *Bulletin of the Institute of Physics* a wide-ranging criticism of both special and general relativity (Brown 1967). One of his criticisms of special relativity was very similar to Herbert Dingle's, namely that the theory required each of two clocks in uniform relative motion to work slower than the other.

The *Bulletin* published an anonymous note on Brown's article in the January 1968 issue (p. 22). The note started by saying that Brown's article had led to a "most voluminous and at times acrimonious correspondence," much of which was critical of the views expressed by Brown. Although the note was anonymous, it included the following statement: "The help of Professor H. Bondi, F.R.S., and his colleagues and Professor R. E. Peierls, F.R.S. in preparing the note is gratefully acknowledged."

After presenting a rebuttal of some of Brown's arguments, the note concluded with the following paragraph:

> It is regretted that it will not be possible to continue discussion of the subject of Dr Burniston Brown's article in the columns of *Physics Bulletin*.

The note, with its closing of the discussion, is a blatant suppression of criticism and appeal to authority. Brown told me in a letter that he was not given the opportunity to reply.

Why is criticism missing in science? Perhaps it is because of the type of education that scientists receive, as described, for example, by Feyerabend (1981):

> Scientific 'facts' are taught at a very early age and in the very same manner in which religious 'facts' were taught only a century ago. There is no attempt to waken the critical abilities of the pupil so that he may be able to see things in perspective. At the universities the situation is even worse, for indoctrination is here carried out in a much more systematic manner. Criticism is not entirely absent. Society, for example, and its institutions, are criticized most severely and often most unfairly and this already at the elementary school level. But science is excepted from this criticism. In society at large the judgement of the scientist is received with the same reverence as the judgement of bishops and cardinals was accepted not too long ago.

Feyerabend also wrote that "science has now become as oppressive as the ideologies it had once to fight. ... Heretics in science are still made to suffer from the most severe sanctions this relatively tolerant civilization has to offer." This is not a new phenomenon: Alfred O'Rahilly wrote in 1940 that "I have discovered that there is a much stricter orthodoxy in science than in theology." (Gaughan, 1986, 182) O'Rahilly was very familiar with both fields: he became a Roman Catholic priest later in life, after the death of his wife.

M. J. Mulkay (1972) commented as follows on the rigidity with which the sciences present their material and the strict intellectual conformity that they demand of their students:

> When students are not being taught established facts and associated theories they are learning how to get the correct answers to precise problems set in terms of the current framework. Facts, theories and problems are typically contained in textbooks which present each body of material from a single, unquestioned perspective.

Although my original university education was in electrical engineering, when I decided to become more active in studying the relativity controversy I started to take courses in philosophy and logic in the University of Toronto, and eventually I received in 1991 a Bachelor of Arts degree with a major in philosophy. I discovered that the study of philosophy is quite different from the study of physics or engineering as described by Feyerabend and Mulkay. I came to realize that the study of philosophy is not just the study of established ideas, or learning how to give the "right" answers in examinations: in order to study philosophy the student must philosophize; he or she must learn to criticize the ideas of others, even the ideas of the most eminent philosophers and thinkers. At the same time, students must be willing to abandon ideas that they have considered incontrovertible. A philosopher who comes to realize that a long-held opinion needs to be corrected or abandoned has made progress. For example. Here is what the editor of one of the philosophical anthologies I studied wrote in his preface (Kaufmann, 1968):

> One of the first things you discover in the study of philosophy is that intelligent and even brilliant men are frequently in disagreement, and that some of the most interesting problems seem to have no ready-made solutions. It might be argued that anyone who has not found this out still needs an education.

The writer went on to say that many a student finds this discovery uncomfortable, and "would like to learn, once and for all, the 'central doctrines' of the great philosophers: that way he has something definite that he can memorize, and does not have to think, himself." He also said that "It is part of the point of philosophy to make men a little less sure about things."

I believe that it is time for physicists to become a little less sure of themselves and what they know, and to accept the fact that criticism of their beliefs should be encouraged rather than being suppressed. In the next chapter I describe some of the difficulties in publishing criticism of relativity.

Chapter 15

Censorship in Science

> If all mankind minus one, were of one opinion, and only one person were of the contrary opinion, mankind would be no more justified in silencing that one person, than he, if he had the power, would be justified in silencing mankind.
>
> John Stuart Mill (1874).

It is well known that history is written by the winners, and this is indeed true in science. In some cases those who win in science do so by the superiority of their arguments, and in some cases they win by simply outliving their opponents. The second method is more common than the first, judging the frequently quoted remark by Max Planck (1949): "A new scientific truth does not triumph by convincing its opponents and making them see the light, but rather because its opponents eventually die, and a new generation grows up that is familiar with it."

There is, however, another important method by which the winners in science achieve and preserve their winning status; that is by the censorship of inconvenient truths. Anyone who has read Herbert Dingle's *Science at the Crossroads* has seen plenty of evidence that such censorship occurs, but I think it may be relevant to present some examples of my own experience of this phenomenon.

In October 1973 I sent a critical paper entitled "Einstein's Special Theory of Relativity" to Dr. P. R. Wallace, Editor of the *Canadian Journal of Physics*. I received an acknowledgment dated October 12, 1973, signed by a secretary, which said in part: "As soon as I have heard from the referee, I shall let you know." A few days later I received the paper back with a covering letter dated October 16, 1973, signed by Dr. Wallace, which read as follows:

> We do not believe that the enclosed paper is of such a character as to be suitable for publication in the *Canadian Journal of Physics*. We suggest that you consider as more appropriate a journal of the character of *American Journal of Physics* or *Nature*.

No mention was made of any referee. I do not think that Wallace really expected that the journals he mentioned would be any more receptive to my article. One feature of this situation that I find personally annoying is the fact that the letterhead of the *Canadian Journal of Physics*, used by Dr. Wallace for his letter of rejection, carried insignia representing the Government of Canada and the statement "Published by the National Research Council of Canada." In other words, the rejection was financially supported by my own taxes, and in particular by that portion of my taxes that is supposed to contribute to the advancement of knowledge.

In 1976 I sent an article to the editor of *Science* with the following covering letter dated August 4, 1976:

> I enclose an original and two copies of an article entitled "The Dingle Affair: An unresolved scientific controversy"; will you please consider this for possible publication in Science.
>
> This article is a factual account of some of the treatment that Professor Dingle's criticism of special relativity has received at the hands of the scientific community, and I believe it would be of interest to your readers.

I received the manuscript back, accompanied by the following letter from the Editor, Dr. P. H. Abelson:

> Thank you for letting us see your manuscript entitled "The Dingle Affair." Unfortunately, we now have a substantial backlog of accepted articles, and we are obligated to give them first priority for publication. Hence, we cannot handle your article at this time.
>
> Accordingly, I am returning your manuscript.

In assessing the cogency of Dr. Abelson's reason for refusing to publish my article, it may be noted that the returned manuscript was date-stamped Aug. 6 1976, presumably the date on which it was received by the journal, while his letter of rejection was dated 9 September. It seems odd that it should take a weekly journal more than a month to realize that it had a "substantial backlog of accepted articles." Obviously any article could be rejected by such a formula; the reason given is obviously unrelated to the merits of the article or the truth of its factual statements.

In May 1981 I sent a copy of a paper entitled "Some Problems in Special Relativity" to The Franklin Institute for consideration for publication in the Institute's *Journal*. I was encouraged to hope that my paper would receive favourable consideration by the fact that the following statement appeared in issues of the *Journal* showing its policy:

> Papers will be considered in all traditional branches of mathematics and the physical sciences, pure and applied, as well as in interdisciplinary fields or composite sciences that combine the philosophies of two or more disciplines. No technical field to which important contributions are being made is excluded—whether experimental or theoretical, mathematical or descriptive.

This appearance of receptivity was enhanced by a further statement that "Articles of general interest will be given special attention and consideration."

The following is the text of the letter that I received in reply, dated June 5, 1981, signed by the Managing Editor, Mrs. L. H. Falgie:

> I am returning your manuscript entitled "Some Problems in Special Relativity" because it unfortunately is not suited as a contribution to the *Journal of the Franklin Institute*. It is more appropriate as a contribution to such publications as *Nature*, *Science*, *Physics Today*, *New Scientist*, *etc.*., rather than this journal.
>
> However, thank you for sending it to us for consideration. .

The above examples are just a selection from my files, and I think that they are sufficient to show that criticisms of special relativity are treated very unkindly by scientific journals. I am glad to be able to report that the article that was rejected by the *Journal of the Franklin Institute* was the one that appears in Appendix II (McCausland, 1983b).

I will now describe a more recent rejection of a paper that I wrote to commemorate the thirtieth anniversary of Herbert Dingle's death. In January 2008 I wrote an article entitled "A Question of Relativity," which presented part of the story of Herbert Dingle's attempt to obtain answers to his arguments about the special theory, and which pointed out the ineptitude of some of the published attempts to answer the question he had asked in his book. I discussed two kinds of irrelevant answers to Dingle: attempts to save the special theory by invoking the general theory of relativity, and attempts to disprove the existence of an internal contradiction by citing experimental evidence. I thought that it would be appropriate to publish such an article to mark the thirtieth anniversary of Dingle's death, and I thought that the passage of time would have blurred the memories of Dingle's campaign sufficiently that the article would not appear to be too controversial.

There were two journals that seemed to me to be suitable for publication of my article. One was *Studies in History and Philosophy of Science*, the journal that had published Chang's article (1993), which had been devoted mainly to Dingle's earlier controversy on the subject of the clock or twin paradox. I thought that the journal might be willing to publish a subsequent article on his later controversy on the subject of the claimed contradiction in the special theory.

There was, however, another journal that I thought would be even more suitable for my article, namely the *British Journal for the History of Science*. I knew that Professor Dingle had been a founding member of the Society that publishes the Journal, and I thought that the Journal would be sympathetic to an article about him. I also thought that the Journal might be glad of an opportunity to atone for the very strange review of *Science at the Crossroads* that it had published (Pyenson, 1976), four years after the appearance of the book, in which it had incorrectly stated that Dingle had died in 1974. Apparently the Journal had not removed Dingle's name from its circulation list before publishing the review, so he was able to read the announcement of his own death, and his reaction may be judged from the following excerpt from his letter to me, dated 1976 November 24:

> A curious incident has just occurred. The latest issue of the *Brit. Journ. Hist. Sci.* contains, after more than four years, a brief "review," if it can be so called, of my book which says that "the late Professor Dingle" "until his death in 1974" believed himself "the only one with courage enough" in effect to believe that s.r. was fallacious; it then states, dogmatically, his "misperception." The writer was "Lewis Pyenson," a name unknown to me. I have sent in a letter for publication, merely poking fun at it, but have had private correspondence with the editors (a change of editor is now occurring

> quite independently of this) who are most profuse in their apologies and promises of amends in the next (March) issue. The gaffe is all the more incredible since I was the virtual founder of the Society that runs the journal—I summoned and presided at the meeting some 30 years ago at which it was formed—and took a leading part in its beginnings. Those now running its journal don't seem to know their own history. I suspect that they invited more noted people to review the book, who refused since they couldn't answer it, but of course the editors can't disclose this to me.

Dingle's published reply to the review can be found in the March 1977 issue of the Journal (Dingle, 1977a). Although he treated the matter lightheartedly, one can imagine his feelings at being treated like that by the Journal that he had done so much to establish.

It was because of Dingle's earlier association with that Journal that I decided to send my paper to it. I did not mention the Journal's review in my article, even though Chang had mentioned it in his article (with less justification, since his article was about the clock paradox controversy and not about the later controversy covered by *Science at the Crossroads*). Nor did I mention the review when I communicated my paper to the Journal, since I did not expect that the Journal's earlier gaffe would require it to confer any advantage on my paper in considering it for publication.

When I had finished writing the article in mid-January 2008, I thought that it might have a reasonable expectation of getting into print around the time of the 30th anniversary of Dingle's death on 4 September 1978. On January 20 I sent it by email to the editor, Professor Simon Schaffer, Professor of the History of Science at Cambridge University. After a period of 2 hours and 9 minutes, according to my email records, I received a reply from Professor Schaffer saying in part: "Thanks for your paper on Dingle and relativity. I regret that it has not been judged appropriate for this Journal." He went on to suggest that I consider sending it to a journal such as *Studies in History and Philosophy of Science*. No reason of any kind was given for the article to have been judged inappropriate, or for the possibility that the other journal might be a more suitable vehicle for it. It is interesting that the advancement of science and technology has now given editors the ability to reject undesired manuscripts at the speed of light.

A like-minded colleague, to whom I sent a copy of my article and a description of the rejection, expressed to me his shock that an article about someone I had described in the first sentence of my article as "one of the founding members of the British Society for the History of Science" had been treated like that. I told my colleague that my description of Dingle was in fact exceedingly modest, and that he was essentially the founder of the Society. I think that Dingle himself would have been extremely shocked if he had known that the Journal run by the Society he was instrumental in founding would reject such a paper without any reason being given.

After that rejection I decided to send my paper to the other journal, the title of which is now *Studies in History and Philosophy of Modern Physics*. I changed the wording of the first paragraph of the paper, since it was no longer relevant to mention that Dingle had been a founding member of the British Society for the History of Science. About two weeks after electronic submission of the article I received the following email message from the Managing Editor of the journal:

> Thank you for the submission of "A Question of Relativity" to *Studies in History and Philosophy of Modern Physics*.
>
> I regret to inform you that the editors decided that we cannot accept the paper for publication, since it is not sufficiently in contact with the current literature on the topic addressed. In particular, we would like to mention the book by Leslie Marder, "Time and the Space Traveller" which discusses the issues raised in your paper but is missing from your reference list.

The reason given for rejection is bizarre. Marder's book did not discuss the issues raised in my paper, since it was published in 1971, before Dingle's *Science at the Crossroads* appeared in 1972, whereas my article dealt entirely with matters that arose from the publication of Dingle's book. In fact, if I had chosen to include Marder's book in my reference list, I could have included in my article another inept attempt to refute Dingle, by Nobel Laureate Professor Max Born, about whom Marder wrote (Marder, 1971, 144):

> In a sense, it was a pity that Born then took up the challenge, because a satisfactory answer to Dingle needed more time than Born wished to devote to the matter. His brief reply, in *Nature*, consisted largely of a 'correction' to Dingle's question (hardly likely to produce the desired effect) and a partially explained space-time diagram.

After receiving both these rejections, I placed my article in the April 2008 issue of the online journal *Apeiron* (McCausland, 2008);

it is reprinted in Appendix VII. As readers can easily verify, nine of the twelve references in the paper are more recent than Marder's 1971 book, which demonstrates the utter absurdity of the reason given for the second rejection of the paper.

Although, knowing from experience the futility of such action, I did not at first dispute the reasons or lack of reasons given by the editors of the above two journals, some months later I wrote the story of those two rejections, roughly in the form in which it appears above, and sent it to both editors and to the President of the British Society for the History of Science. Dr. Jeff Hughes, President of the Society, was kind enough to write to me about it, and I would like to comment on what he said.

Dr. Hughes wrote that his own view was that the article was just too narrowly focused and insufficiently contextual to warrant publication in *BJHS*. Had he been a referee, he would not have been able to recommend publication. In reply, I conceded that my article was narrowly focused, but said that if I had had a constructive criticism of the paper, along the lines of his comments, I might have been able to write a more wide-ranging article that would have met the criticism. (I should also add that, in writing a paper that might be controversial, I think it best to concentrate on a small number of sound arguments rather than including other arguments that might be more vulnerable.)

Dr. Hughes made the following comment about the rejection of my paper by *BJHS*:

> Since reading your paper I have discussed the matter with the Editor of BJHS, and can confirm that the rejection of the paper was entirely on grounds of its limited historicity—immediately apparent to any experienced historian—rather than its intellectual content *per se*. Your characterisation of its rejection as 'suppression' or 'censorship' is therefore both inaccurate and unjustified.

I replied to that point as follows:

> You tell me that the Editor of *BJHS* rejected my paper on grounds of its limited historicity, rather than its intellectual content, so my "characterisation of its rejection as 'suppression' or 'censorship' is *therefore* both inaccurate and unjustified." But the point is that the reason was not stated, so my characterisation was based on the evidence that I had before me, which obviously did not include that reason. The fact that there exists in the mind of an Editor a valid reason for not accepting a paper is not obvious to the recipient of a flat

> rejection, if the reason for the rejection is not stated. Justice should be seen to be done.

In my reply to Dr. Hughes I also re-stated my view that the other editor's rejection was bizarre. I am sorry that I am unable to identify by name the scientist who was responsible for the bizarre rejection, who made no comment to me on the matter.

Although censorship of inconvenient truths may be an effective temporary way of maintaining the appearance of harmony and unity in the scientific world, it is unlikely that such suppression will end the disputes over special relativity. Some day, for example, the scientific community may begin to notice the ineptitude of the arguments that were used to answer Herbert Dingle's criticisms, and wonder why those arguments were accepted by almost everyone as being sound. Anyone who reads Herbert Dingle's story in *Science at the Crossroads* will know that suppression of his arguments by scientific journals was an important part of the process, and that the actions of journal editors eventually become part of the historical story. Although the journals I have discussed above are not so important that their treatment of my papers would change the course of scientific history, perhaps my account of those rejections may provide an interesting footnote to the historical record.

The censorship of criticisms of special relativity continues, as described by Hans J. Zweig (2005). The following is an excerpt from an email message to Zweig from the secretary of the Editor-in-Chief of the journal *Annalen der Physik*, the journal in which Einstein's original paper on the special theory appeared in 1905:

> On behalf of Professor Eckern, Editor in Chief, *Annalen der Physik*, I acknowledge receipt of your fax ... relating to your paper "Einstein's Fuzzy Logic."
>
> ... there is no hope of getting this material published in *Annalen der Physik*, particularly as the editors recently unanimously decided not to publish, or to even discuss contributions which claim to have obtained counter-examples to Einstein's Theory of Special Relativity ...

It is not just criticisms of the original papers on relativity that are censored. Moffatt (2010) has described how he found serious mistakes in a paper by Einstein and Infeld (1949). He and a colleague felt obliged to point out the mistakes, and submitted a paper to *Physical Review*, which he described as "the premier physics journal in America." He described the journal's reply as follows (p. 88):

> The letter stated that *Physical Review* could not publish our paper because it would 'besmirch the reputation' of one of the greatest physicists of the twentieth century. I was so upset by this intellectual dishonesty that I contemplated leaving physics.

As Moffatt pointed out, the paper that he and his co-author had criticized had been so highly regarded by the editors of the *Canadian Journal of Mathematics* that "its first page consisted of the portrait of Einstein taken by the famous Canadian photographer Yousuf Karsh." It seems to me that excessive hero-worship of Albert Einstein may have clouded the judgment of those who were involved in assessing the paper.

Epilogue

> If I have a religion, its first precept is that we shall seek truth faithfully; and I would say with Emerson: "God offers to every mind its choice between truth and repose. Take which you please. You can never have both."
>
> William McDougall (1927)

> It is said that it is better to travel hopefully than to arrive, and the dogma that we have arrived at the truth is not conducive to further mental effort.
>
> E. G. Cullwick (1959)

> ...the belief that there is only one truth and that oneself is in possession of it, seems to me the deepest root of all that is evil in the world.
>
> Max Born (1968)

Herbert Dingle's criticisms of special relativity were based on his love of science. In *Science at the Crossroads* he repeatedly referred to a quotation from a book by Sir Henry Dale, a former President of the Royal Society, which extolled the virtues of science in its "patient devotion to the search for objective truth." On page 116 Dingle gave this explanation for the choice of the title of his book:

> I have called this book *Science at the Crossroads* because science has indeed now two courses before it: it can, on the one hand, resume progress along the course which Dale,

> generalising from its past history, believed it still to be following, or it can, on the other hand, remain on the path which I hope I have enabled the reader to see that it is now pursuing. The consequences of the choice are important beyond measure.

Herbert Dingle was trying to save science from itself, but failed. It is still possible, however, for science to redeem itself by having another look at the relevant arguments. Perhaps someone in a position of influence may notice the ineptitude of the attempts that were made by many people to answer his criticisms of special relativity, and may consider seriously the possibility that they may have been wrong and Dingle right. Since the defenders of the theory said many inconsistent things, it is obvious that they could not all have been right.

Who could possibly be a candidate for such a role? Perhaps some Editor of *Nature* might know enough simple logic to realize that some of the supposed refutations of Dingle's arguments that appeared in *Nature* were highly illogical, and might decide to do something about it.

Perhaps some Nobel laureate might come to notice the ineptitude of the collection of contradictory arguments that have been generally accepted as refutation of Herbert Dingle's criticisms. Unfortunately the four Nobel prizewinners in my story have clouded the issue rather than clarified it. Max Born was so confident in being in possession of the one and only truth that he refused even to read Herbert Dingle's reply to his argument, claiming that his own argument was irrefutable. P. A. M. Dirac wrote that modern experimental results support Lorentz's theory rather than Einstein's, but gave as his reason for belief in special relativity that it puts importance on the Lorentz transformations which are beautiful mathematically. Willis Lamb listened politely to a critic of relativity until he mentioned Herbert Dingle, whereupon he went into a purple rage and ordered the critic to leave his office. P. M. S. Blackett taught special relativity for many years, but was unable to understand Dingle's criticisms and seemed unwilling to take the trouble to try to understand them.

Part of my own motivation for continuing my scientific adventure for so long has been to try to draw attention to the injustice that was done to Herbert Dingle by the scientific community. Among other things, I have pointed out the complete irrelevance of experimental support for the special theory in attempting to refute a claim that the theory is self-contradictory. The only way in which the question of a self-contradiction in the special theory can be answered is by examining the theory itself, in the form in which it was published, not by examining other properties of the world, and especially not by examining the general theory. This shows the irrelevance of the arguments used by many scientists to try to refute Dingle's thesis.

In the spirit of my epigraphs to this chapter, I do not claim to be in possession of a unique truth, and I believe we should be guided by T. H. Huxley's (1894) statement that "the scientific spirit is of more value than its products, and irrationally held truths may be more harmful than reasoned errors." I have documented several cases in which physicists have claimed that experimental support for the special theory refutes Dingle's claim that there is an internal inconsistency in the theory, and I have shown that those claims are irrational. The problem is that the physics community is so completely certain that the special theory is sound that influential physicists refuse to discuss it or to allow publication of informed criticism, and do everything they can to foster the belief that critics of special relativity are crackpots.

If Herbert Dingle was wrong in claiming that there is an inconsistency in the special theory, it is reasonable to expect that he would have been proved wrong by a clear argument instead of the collection of self-evident contradictions that have been put forward in answer to his criticisms by the defenders of the theory. Just as Dingle wrote that resolution of the paradox required both proof of one of the two inconsistent outcomes and disproof of the other, in a similar way a sound refutation of his argument should include a public repudiation of those defences of the theory that contradict that refutation. The existence, and the continued acceptance, of the mutually inconsistent set of defences in the literature is a blot on the escutcheon of science.

There are at least two arguments against the special theory that, to the best of my knowledge, have never been refuted. One is Dingle's argument in an appendix of *A Threefold Cord* (Samuel and Dingle, 1961), which I have reproduced in my Appendix I with an added commentary, and the other is my own argument that appears in Appendix IV. As I have shown in my commentary in Appendix I, the distinguished mathematician Professor I. J. Good tried to refute Dingle's argument but failed. Professor Good, who was also the only person who has engaged in significant public debate with me on the subject, ended by admitting a serious error in his own argument and apologizing to me.

In 1972, science was at the crossroads. Today, forty years later, science still has the choice between truth and repose, but it can never have both. In their attitude to special relativity, scientists are so deeply in repose that it seems a pity to rouse them from their dogmatic slumber, yet I do not think they should be allowed to rest in peace.

APPENDICES

Appendix I

Dingle's Argument

In Appendix II of A *Threefold Cord* (Dingle and Samuel, 1961), Dingle presented one of his earliest arguments claiming that special relativity could not correspond with fact. The following is most of the text of that Appendix; it comprises that part of the Appendix that was included in a manuscript that had earlier been submitted to *Nature* but was not published there. In order to simplify the discussion in my Commentary at the end of the present Appendix, I have made a trivial change in the notation, denoting the quantity $1/\alpha$ by γ in accordance with more common usage.

Proof that Einstein's Special Theory cannot correspond with fact*

Consider a group of bodies all relatively at rest. Choose one of them as the origin of co-ordinates, and place a standard clock there. Then every one of the bodies is at rest in a co-ordinate system thus chosen, and the time of any event that occurs among them is obtained by letting a beam of light proceed from the event to the clock at the origin and subtracting from the time which the clock records at its arrival the quantity r/c, where r is the distance of the event from the origin and c is the velocity of light. This is the definition adopted by Einstein, and there is no reason to quarrel with it.

* From: A Threefold Cord: A discussion between Viscount Samuel and Herbert Dingle, by Herbert Louis Samuel, Viscount Samuel, Copyright 1961, Allen & Unwin. Reproduced by permission of Taylor & Francis Books UK.

It is often expressed in a slightly different form. We suppose a standard clock, at rest in the co-ordinate system, placed at each point of space and synchronized with the clock at the origin by the dispatch of a light signal from the origin to this clock and its immediate return. Then the clock in question is set so that it records the time of receipt of this signal as the mean of the times, by the clock at the origin, of its dispatch and return. The time of any event is then given directly by the reading of the clock at the place and time at which the event occurs.

It is obvious that this gives the same value for the time of the event as the former definition, and indeed the same value is given by any clock of this set, if we make the proper allowance for the time of travel of light from the event to that clock. In other words, the time of an event in any co-ordinate system is a property of the *system*, not of any single clock in it. Though this will hardly be questioned at this stage it is necessary to insist on it because many errors have been made in applying the Lorentz transformation (which so far we have not introduced) to actual events, which can be traced to a tacit assumption that the time of an event in a single co-ordinate system varies with the clock, stationary in that system, by which it is determined.

Einstein was perfectly clear on this point. In his original paper on the subject[†] he wrote: 'It is essential to have time defined by means of stationary clocks in the stationary system, and the time now defined being appropriate to the stationary system we call it "the time of the stationary system" '. It is obvious from this that the time, in any single co-ordinate system, of any event has a single value, no matter by what clock of the system it is evaluated. Again, Einstein and Infeld, after giving the definition of time in a co-ordinate system in terms of a multitude of clocks as described above, write[‡]: 'When discussing measurements in classical mechanics, we used one clock for all co-ordinate systems. Here we have many clocks in each co-ordinate system. This difference is unimportant. One clock was sufficient, but nobody could object to the use of many, so long as they behave as decent synchronized docks should.'

---

† *Ann. d. Phys.*, 17, 891 (1905); English translation in *The Principle of Relativity* (Methuen, 1923), p. 40.

‡ *The Evolution of Physics* (Cambridge, 1938), p. 191.

It is clear, then, that one clock is sufficient, and we shall suppose that that is all that we have—a single clock at the origin that gives the standard time for the system of every event that occurs.

Next, it must be noticed that we say 'the time for the system' and 'every event that occurs', not 'every event that occurs in the system'. An event—a point-instant—is independent of all coordinate systems; it is meaningless to speak of it as 'stationary' or 'moving' in any such system. If the event is the instantaneous collision of a large number of bodies, all coming from different directions, it can be regarded as happening to any one of them: they are all moving differently, but the time of the event in any co-ordinate system is exactly the same, no matter on which of the bodies we regard it as occurring. Events are essentially independent of co-ordinate systems, but their places and times may vary with the system chosen.

Suppose now that we have two groups of bodies, all the bodies in each group being relatively stationary and moving with uniform velocity with respect to all the bodies in the other group. We may think of two swarms of stars passing through each other, like the two star streams of Kapteyn, the peculiar motions of the stars in each stream being neglected. Suppose that the stars at the origins pass one another at an instant at which the standard clocks placed on them both read zero, and that at some later time two stars, one in each group, collide. That is an event, and it has a definite time in each co-ordinate system. It is the same event, no matter whether the observer at the origin of stream one regards it as a disturbance of one of his stationary stars or whether the observer at the origin of stream two regards it as a disturbance of one of *his* stationary stars, but the times of the event which the observers record may of course not be the same.

Let us suppose they are not. Then since the clocks agreed when they were together, they must have run at different rates afterwards, for if we want to compare the rates of clocks the only conceivable way of doing so is to compare the time intervals which they record between the same events. The events belong no more to one co-ordinate system than to the other, but the times that elapse between them are properties of the systems, *i.e.*, of the clocks that record the times of the systems. Hence, if one clock shows, say, twice the time interval shown by the other between the same two events, that clock must go at twice the rate of the other.

Now the clocks are supposed to be of identical construction, so that if one is regarded as running uniformly in its system, the other also must be regarded as running uniformly in *its* system. Hence, if the *clock-time* intervals in the two co-ordinate systems, between the same two events, are ΔT and $\Delta T'$, respectively, $\Delta T/\Delta T'$ must be a constant quantity; it cannot vary with the particular events chosen for the comparison.

From this it follows inevitably that the co-ordinate differences, Δt and $\Delta t'$, to which the Lorentz transformation applies, cannot, as Einstein's theory requires, represent the clock-time intervals, in the respective co-ordinate systems, between the same events. For, according to the Lorentz transformation,

$$\Delta t'/\Delta t = \gamma - (\gamma v/c^2)(\Delta x/\Delta t)$$

where $\gamma = (1 - v^2/c^2)^{-1/2}$. We therefore get different values for this ratio according to the events we choose, Δx being the space interval and Δt the co-ordinate time-interval, in one of the systems, between those events. We can, by a suitable choice of events, make $\Delta t'$ greater or less than Δt, by any amount we like. But, as we have seen, the relative rates of the clocks must be independent of the events chosen for their determination. Hence $\Delta t'/\Delta t$ cannot represent the relative rates of the clocks.

Commentary

As far as I am aware, none of Dingle's critics attempted to publish a rebuttal of this argument when it was first published. More recently, however, Good (2008) attempted to refute the argument in his review of *Old Physics for New* (Phipps, 2006) in which Phipps also included the part of Dingle's appendix that I give above.

I think Good misinterpreted the last paragraph of Dingle's argument. In his equation, Dingle intended $(\Delta x,0,0)$ to represent the spatial separation vector and Δt the time interval between the two events that he mentioned, namely the meeting of the origins of the two systems and the collision of the two stars. Good wrote as follows:

> But actually both sides of the equation are equal to $(1 - v^2/c^2)^{1/2}$ whatever the value of v and whatever events we choose, assuming of course that we are referring to a pair of clocks for which the one has velocity v relative to the other. Dingle seems to be saying that the right side of [the equation] can have various values while the left side remains fixed. I think Dingle made a mistake here.

Dingle did indeed refer to a pair of clocks for which one is assumed to have velocity v relative to the other, but when Good argued that both sides of the equation are equal to $(1 - v^2/c^2)^{1/2}$ he was taking $\Delta x/\Delta t$ to be v, and that is a completely different assumption. Having taken one event to be the meeting of the origins of the two systems, the only other events for which $\Delta x/\Delta t$ is equal to v are those that occur at the origin of the second system, where $x' = 0$ and where one of the clocks is located; in other words, the clock at $x' = 0$ is present at both events. This makes it difficult to see how this is valid for "whatever events we choose," as Good claims it is. It is clear from Dingle's argument that, if we have only one clock in each system and we wish to compare the clock intervals in the two systems between a pair of *arbitrary* events, then it is necessary to take into account the times for light to travel from *both* events to *both* clocks. By taking $\Delta x/\Delta t$ to be v, Good has entirely neglected the possibility that the two events can be chosen arbitrarily, so that $\Delta x/\Delta t$ can have an arbitrary value. It is this latter, perfectly allowable, case that Dingle has assumed in his argument.

Appendix II

Problems in Special Relativity*

Arguments that have been used to defend the special theory of relativity against criticism contain many inconsistencies. These problems should be thoroughly and objectively examined by scientists and philosophers to attempt to ascertain the truth of the matter.

Ever since Einstein's special theory of relativity became a prominent part of physics, it has been a subject of some controversy. One of the foremost critics of the theory was the late Herbert Dingle (1890-1978), who spent much of his time and energy during the last two decades of his life in trying to persuade the scientific world that the special theory, although mathematically valid, contains an inconsistency in its physical application. Although most scientists seem to be convinced that the controversy stirred up by Professor Dingle's criticisms has been conclusively settled in favour the theory, a close examination of the relevant literature shows many inconsistencies in the arguments by which the special theory has been defended. The present article does not attempt to settle the matter, in fact it attempts to show that the issue has *not* yet been satisfactorily settled. It is hoped that scientists and philosophers may be encouraged to continue the search for the truth of the matter, whatever it may be.

* Reprinted from *Wireless World*, Vol. 89, No. 1573, pp. 63-65, by permission of *Electronics World*.

Simple example of inconsistency

Readers who are not experts on relativity may feel convinced that the inconsistencies that have been mentioned are beyond their understanding; on the contrary, many of them are perfectly obvious to anyone who takes the trouble to read them. To take a specific example, consider two inconsistent statements that were made in the British journal *The Listener* in 1971.

Professor J. Taylor claimed[1] that the results of the well-known experiment of Hafele and Keating, which had then been recently performed, supported Einstein's special theory. Professor Dingle published a letter rebutting Taylor's article, and further correspondence continued to be published, in the course of which another scientist, Professor M. A. Jaswon, published a letter[2] which disagreed with some of Dingle's points, but which agreed with Dingle that the experiment in question had "no relevance whatever for the special theory." Although that statement was directly contrary to Taylor's claim, Taylor later published another letter[3] which continued to criticize Dingle but which took no notice whatever of Jaswon's statement.

Some observers of the controversy may believe that inconsistent statements like these result from attempting to express abstruse technical matters in simple language, and that such inconsistencies may therefore be dismissed as being inconsequential. But the inconsistency between the statements mentioned above cannot be dismissed in that way. A statement that the results of a particular experiment support a certain theory is a perfectly simple factual statement (however abstruse may be the reasoning by which that statement might be justified), and the same applies to the contrary statement. The fact that Taylor's and Jaswon's statements are contrary to one another (that is, they cannot both be true, though they could both be false) shows that, unless there is an inconsistency in the special theory itself, one or other of the two scientists (or both) misunderstood either the theory or the experiment (or both).

It will also be clear to any reader, scientist or not, who reads the whole of the correspondence that includes the above items (refs 1-3), that no attempt was made to resolve the inconsistency between Taylor's and Jaswon's statements. If science is the search for truth, wherever the search may lead, the serious inconsistency between the statements of the two scientists ought to be followed up to find out which statement, if either, is true. The fact that both statements have been accepted in spite of their obvious incompatibility is evidence that there is not enough scientific curiosity about the truth of the matter. The remainder of the present paper presents further evidence in support of the same point of view.

Further Examples of Inconsistency

Professor Dingle's criticisms of special relativity are presented at length in his book *Science at the Crossroads*[4], and it is in the published reviews of that book that many of the inconsistent attempts to defend the theory have been made. To study some of these attempts, consider Dingle's crucial question, which is central to his book, and which is worded as follows:

> According to the special relativity theory, as expounded by Einstein in his original paper, two similar regularly-running clocks, A and B, in uniform relative motion must work at different rates. In mathematical terms, the intervals dt and dt', which they record between the same two events are related by the Lorentz transformation, according to which $dt \neq dt'$. Hence one clock must work steadily at a slower rate than the other. The theory, however, provides no indication of which clock that is, and the question inevitably arises: How is the slower-working clock distinguished?

In a review[5] of Dingle's book, Professor J. M. Ziman quoted the above question and then wrote: "This is a perfectly reasonable question to which science should indeed give an answer." Later in his review he gave his own answer, in the following words: "In fact, the answer to Dingle's 'question' is simple: the fastest-working clock between any two events is one that travels between them by free fall." But, as Dingle subsequently pointed out[6], neither of the events need be at either of the clocks concerned. Also, since the question asked for a distinction between two clocks, not for a choice among all possible clocks, Ziman's answer, whether or not it is a true statement, is simply not an answer to the question that was asked.

Dingle also supplemented his question by referring to a specific example in Einstein's original paper on special relativity, in which Einstein had stated that a balance-clock at the equator would work more slowly than an exactly similar clock at one of the poles. Dingle stipulated that any answer to his question should specify what it was that entitled Einstein to conclude, from the special theory, that the equatorial and not the polar clock worked more slowly. Dingle stressed that the special theory did not take any account of possible effects of acceleration, gravitation, or any difference at all between the two clocks except their state of uniform relative motion. It should be strongly emphasized, however, that he did not assert that acceleration and gravitation were *absent* from the situation described by Einstein, but that those phenomena are not dealt with by special relativity, and consequently it is not legitimate to invoke those phenomena to explain what entitled Einstein to conclude *from the special theory* that the equatorial clock worked more slowly.

The attempts to answer this supplementary question show an interesting diversity. In the first place, it is obvious that Ziman's answer, quoted above, does not apply to this situation; after the two clocks are in their positions at the pole and at the equator, there is no event at which both clocks are present, so there is no way in which Ziman's criterion can distinguish between them unless some pair of events is specified.

Consider now some of the other attempts to answer the question about the polar and equatorial clocks. For example, Professor G. J. Whitrow wrote as follows[7]:

> For a supporter of relativity, the essential difference between the two clocks is that relative to the centre of the Earth (which for the purpose concerned can be regarded as the origin of an inertial frame) the clock at the equator describes a circle and so cannot be associated with an inertial frame, whereas the polar clock is at rest and can be associated with an inertial frame for a period of time during which the curvature of the Earth's orbit can be neglected. The time difference mentioned by Einstein can be demonstrated by means of the Minkowski diagram, in which the track of the polar clock will be rectilinear whereas that of the equatorial clock will be curved.

Two comments may be made about this. First, if the equatorial clock is not in an inertial frame, then its motion lies outside the scope of the special theory, which applies only to inertial frames[8]; it is therefore invalid to deduce from the special theory *any* conclusion about the relative rates of the two clocks. Second, the answer raises the equally difficult question of why a clock that moves in a large closed curve is in an inertial frame, while one that moves in a smaller closed curve is not.

Compare Whitrow's answer with the following answer, which is found in an unsigned editorial article in *Nature*[9]:

> It seems now to be accepted that Einstein's original argument was uncharacteristically loose. The point of the illustration is that a clock at the pole of rotation may be taken to be in an inertial frame which is nearly (but not quite) properly defined by the direction of the Earth's motion around the Sun. The clock at the equator is in another. Einstein's lack of clarity concerns the inertial frame of the observer of the two clocks.

This statement implies that the answer to the question about which clock works more slowly depends on the observer. But Einstein's statement clearly implies that the slowing of the equatorial clock is a real effect and not merely an effect of observation, and this is confirmed by the fact that he added a footnote to say that his statement did not apply to pendulum clocks[10]. The answer[9] also states that the equatorial clock is in an inertial frame, and this explicitly contradicts Whitrow[7], who states that it is not.

Another answer to the same question is given by Stadlen[11], who writes:

> But the relative motion involved in this case, being circular, is non-uniform. I submit, therefore, that Einstein was wrong in saying that his prediction followed from the special theory, which deals only with the effects of uniform motion. This is not to say that the prediction was invalid. For Einstein was, intuitively, anticipating his later general theory, according to which the equatorial clock runs slower because of the centripetal force exerted upon it.

This answer is inconsistent with *both* the previous answers, since it disagrees with Whitrow[7] about whether the result follows from the special theory, and it disagrees with the *Nature* editorial[9] about whether the slower working is real or dependent on the motion of the observer. Furthermore, the fact that the prediction follows from the general theory does not make Einstein's prediction *from the special theory* valid, as Stadlen implies it does. As is well known to logicians, the fact that the conclusion of an argument is true does not guarantee that that argument is valid.

Another interesting attempt to identify a false step in one of Dingle's arguments was made by McCrea[12], who wrote:

> The false step is that Dingle regards the situation treated by relativity as the symmetric comparison of one single clock with another identical single clock (in relative motion). This is not the situation. Actually many colleagues have pointed this out, or given an equivalent answer.

Unfortunately McCrea does not identify any of the "many colleagues" whom he claims to support his argument, but it is clear from the foregoing that Ziman, for example, does not. Ziman states[5] that Dingle's question is perfectly reasonable, and the question, as he correctly quoted it, includes a statement that if there are two clocks in uniform relative motion, the special theory requires one to work steadily at a slower rate than the other. McCrea's statement is also inconsistent with Einstein's statement that a (single) clock at the equator would work more slowly than an exactly similar (single) clock at one of the poles.

Other illogical arguments

In addition to the inconsistencies already mentioned, some of the arguments used in defending special relativity are lacking in logical rigour. To illustrate this, consider some examples.

In one of the earliest attempts to refute Dingle's criticisms, Born[13] wrote as follows:

> The simple fact that all relations between space co-ordinates and time expressed by the Lorentz transformation can be represented geometrically by Minkowski diagrams should suffice to show that there can be no logical contradiction in the theory.

As the Lorentz transformation is contained in the special theory, but is not the whole theory, it is not logically valid to claim that some property of the Lorentz transformation is a sufficient condition for the whole theory to be free of logical contradiction.

In another attempt to refute Dingle, Professor I. Roxburgh[14] discusses Dingle's argument that if there are two clocks A and B in uniform relative notion, the special theory requires A to work faster than B and B to work faster than A, and this makes the theory internally inconsistent. Roxburgh states that Dingle does not even discuss what he means by "faster," and then goes on to say:

> Secondly, why is it impossible for A to go faster than B and B to go faster than A? This depends on the definition of faster. To illustrate this, consider the following two statements:
>
> The Moon is bigger than the Sun.
>
> The Sun is bigger than the Moon.
>
> Are these statements mutually contradictory? This depends on the meaning of bigger. For terrestrial beings the first

> statement is true, for Martians the second is true. The relative size depends upon the position of the observer. So it is with time and clocks."

If it is important to define "faster," it is also important to use other words precisely; yet it is clear from the quotation that Roxburgh does not literally mean "is" in the two contrasted statements, but something like "appears to be." Thus, the two contrasted statements are not analogous to the two statements that Dingle claims to be inconsistent. Or, if Roxburgh does mean the pair of contrasted statements to be taken literally, then he, as a terrestrial being, is asserting that the moon is bigger than the sun. Although we are terrestrial beings, we know that the sun is bigger than the moon, and we know it from observations that have been made from the earth.

To put the matter in terms of logical relations, the expression "is bigger than" represents an asymmetrical relation, whereas Roxburgh's pair of contrasted statements asserts that "is bigger than" is not an asymmetrical relation[15]; there is therefore a contradiction inherent in what Roxburgh has written. Of course, a contradiction between any two statements can be avoided if one is free to disregard literal meanings of words and interpret the meanings of the statements in such a way as to avoid the contradiction. This is similar to the technique described by Dingle (ref 4, page 180) for avoiding the inconsistency in special relativity: "When the theory appears to lead to incompatible objective results, they are written off as merely different *appearances,* but claimed as *realities* when some actual phenomenon has to be explained.

Whitrow has also published an argument[7] which purports to refute Dingle's claim that the special theory is inconsistent in requiring each of two relatively moving clocks to work faster than the other. The last sentence of his argument is:

> Dingle's requirement is therefore equivalent to introducing the Newtonian concept of universal time, and this is incompatible with special relativity.

Now whether or not Whitrow's statement about Newtonian time is true, the sentence quoted does not prove that Dingle is wrong; all it states is that either Dingle is wrong or special relativity is wrong. As the point at issue is the validity of special relativity, and as the context obviously implies that the argument that ends with the quoted sentence proves that Dingle is wrong, Whitrow's argument shows an excellent example of the textbook fallacy known as begging the question[16]. Since Whitrow has subsequently published the same argument two more times[17,18], in obituary notices on Professor Dingle, the pointing out of this logical fallacy is overdue.

The foregoing examples of inconsistencies and logical fallacies in the arguments used to defend special relativity do not in themselves prove that Dingle is right, or that special relativity is wrong. However, if two scientists make inconsistent statements about the same theory, one or other of them must have made an error in deduction, or else the theory itself contains an inconsistency. In other words, the inconsistencies in the statements that have been made by the defenders of the special theory actually support Dingle's case that there is an inconsistency in the theory, rather than refuting it.

Although scientists may be convinced that the conclusion they have already reached is true, they should also be concerned with whether the arguments by which that conclusion has been reached can withstand scrutiny without revealing inconsistencies. I suggest that the scientific ideal towards which science should strive in this case is that stated by T. H. Huxley when he wrote[19] that "the scientific spirit is of more value than its products, and irrationally held truths may be more harmful than reasoned errors." It is time for the truth of this matter to be actively and carefully sought.

References

1. J. Taylor, Views, *The Listener,* vol. 86, 1971, pp. 642-3.
2. M. A. Jaswon, Travelling clocks. *The Listener,* vol. 86, 1971, p. 724.
3. J. Taylor, Travelling clocks. *The Listener,* Vol. 86, 1971, p. 804.
4. H. Dingle, Science at the Crossroads. Martin Brian & O'Keeffe, London, 1972.
5. J. Ziman, Science in an eccentric mirror, *Nature,* vol. 241, 1973, pp. 143-4.
6. H. Dingle, Dingle's question, *Nature,* vol. 242, 1973, p. 423.
7. G. J. Whitrow, Review of "Science at the Crossroads," *British Journal for the Philosophy of Science,* vol. 26, 1975, pp. 358-62, .
8. A. Einstein and L. Infeld, The Evolution of Physics, Cambridge University Press, 1938.
9. "Dingle's answer" (unsigned editorial) *Nature,* vol. 239, 1972, p. 242.
10. A. Einstein, On the electrodynamics of moving bodies, in H. A. Lorentz *et al.,* The Principle of Relativity, Methuen, 1923; Dover, 1952.
11. G. Stadlen, "Dingle's challenge," *The Listener,* vol. 88, 1972, pp. 411-2.
12. W. H. McCrea, Definitions and realities, *The Listener,* vol. 82, 1969, p. 315.
13. M. Born, Special theory of relativity, *Nature,* vol. 197, 1963, p. 1287.

14. I. Roxburgh, Is special relativity right or wrong?, *New Scientist*, Vol. 55, 1972, p. 602.

15. I. M. Copi, *Symbolic Logic*, fifth edition, Macmillan, New York, 1979.

16. I. M. Copi, *Introduction to Logic*, fifth edition, Macmillan, New York, 1978.

17. G. J. Whitrow, "Herbert Dingle," *Nature*, Vol. 277, 1979, pp. 584-5.

18. G. J. Whitrow, "Herbert Dingle," *Quarterly Journal of the Royal Astronomical Society*, vol. 21, 1980, pp. 333-7.

19. T. H. Huxley, The coming of age of 'The origin of species', in *Collected Essays*, vol. 2, Macmillan, 1894.

Appendix III

On Faraday's Ray-Vibrations and the Postulates on the Velocity of Light*

A speculative dissertation by Faraday in 1846 has acquired a renewed significance in connection with recent studies by Moon and Spencer and their colleagues, whose Postulate III appears to be a feasible alternative to Einstein's 1905 postulate on the velocity of light. Objective discussion of the subject is extremely important because of the involvement of special relativity and quantum theory.

Key words: velocity of light, relativity, quantum theory

In 1846 Faraday published a highly speculative paper entitled "Thoughts on Ray-vibrations"[(1)] in which he suggested that every light source had a system of "rays," or lines of force, extending indefinitely in all directions from the source, and that the propagation of light occurred by transverse vibrations on these lines of force.

Faraday's paper has recently acquired a renewed significance, in view of some recent publications of Moon and Spencer and their colleagues[(2-5)], in which they discuss possible postulates on the velocity of light and show that a postulate introduced by Moon and Spencer in 1956[(6)] is a possible alternative to Einstein's postulate on the velocity of light from moving sources in his original paper on

* Reprinted from *Physics Essays*, Vol. 5, No. 3, 1992, pp. 176-7, by permission.

special relativity[7]. In their 1956 paper. Moon and Spencer considered the possibility that light radiation might "remain in some way coupled to its source even after it has been emitted," and this led to their postulate that "the velocity of light at any instant must be constant (c_0) with respect to the source *at that instant.*" [Italics in the original.]

Although Faraday did not state any postulate on the velocity of light in his paper, the ideas that he presented suggest that he had something in mind that was very similar to Moon and Spencer's postulate, and Dingle interpreted his ideas in a similar way when he wrote[8]:

> Faraday once suggested that every atom carried with it a system of "rays" or "lines of force," extending indefinitely in all directions and moving instantaneously with it. The rays from one atom could freely interpenetrate those from others, and they could serve as a vehicle for transmitting light and electromagnetic influences. The ensemble of all such rays in space can be called an ether, but it would be impossible to ascribe to any body a velocity with respect to it. The body must always be at rest with respect to its own rays, and have various velocities with respect to all other sets of relatively moving rays, so that the only measurable velocity would be that of one body with respect to another.

Although one might disagree with Dingle's identifying the ensemble of rays as an ether, whereas Faraday stated that he wished to dismiss the ether but not the vibrations, I believe that the mental picture envisaged by Dingle is a valid representation of Faraday's ideas and shows the similarity between those ideas and the Moon-Spencer postulate. It is not particularly important who first thought of those ideas; what is important is that the ideas be given full and objective consideration on their merits. In this connection, although Moon, Spencer and Moon[3] write that the new postulate introduced in 1956 "has hitherto been largely ignored partially because of the obscure way in which it was stated," a study of the literature shows that Grünbaum published in January 1957 a "Discussion"[9] concerning the 1956 paper, in which he rebutted a comment that Moon and Spencer had made about his interpretation of Einstein's views on simultaneity of events in a single reference frame. Although Grünbaum's Discussion may not have been directly related to the "new postulate," I believe that he had a valid point, but I have been unable to find any response by Moon and

Spencer to his Discussion. In fact, they published another paper in the same journal in October 1958,[10] in which they referred to their 1956 paper[6] but not to Grünbaum's Discussion of it; nor do they mention Grünbaum's Discussion in their recent papers. It seems rather odd that they should now give the impression that their paper was largely ignored, when it was discussed in the same journal in which it was published, apparently without reply.

In their recent papers Moon and Spencer and their colleagues[4,5] have shown that their postulate, modified if necessary by the added qualification that the coordinate system not be in rotation, is consistent with the results of the Michelson-Morley, Michelson-Gale, and Sagnac experiments. In other words, what they now call Postulate III is a possible alternative to Postulate I, which is Einstein's postulate on the velocity of light in his original paper on special relativity. In view of these findings, I would like to support these scientists in their plea[2] for open-minded investigation of the postulates of currently accepted physics. It would be difficult to exaggerate the importance of giving serious consideration to this postulate: what is at stake is the special theory of relativity.

Although there are many scientists who have questioned the validity of the special theory of relativity because of internal problems of the theory, there have arisen in recent years new reasons for questioning the special theory because of its apparent conflict with quantum theory. The apparent conflict is based on theory and experimental results relating to tests performed on pairs of particles which interact and then move apart; correlations between measurements on the two particles appear to prove that some kind of influence can travel instantly from one to the other. Scientists have been aware that there is a possible problem in this area since the publication of a very famous paper by Einstein, Podolsky and Rosen,[11] the significance of which has been debated ever since. One of the most important contributions to the debate is a famous paper by Bell[12]; the following is part of the conclusion of his paper:

> In a theory in which parameters are added to quantum mechanics to determine the results of individual measurements, without changing the statistical predictions, there must be a mechanism whereby the setting of one measuring device can influence the reading of another instrument, however remote. Moreover, the signal involved must propa-

> gate instantaneously, so that such a theory could not be Lorentz invariant.

One important reservation expressed in Bell's paper was the crucial necessity of performing experiments in which the settings of the apparatus were changed during the flight of the particles. Such experiments have now been performed and, in a more recent assessment of the situation, Bell has written as follows,[13] referring to the de Broglie-Bohm picture of quantum phenomena:

> This picture, and indeed, I think, any sharp formulation of quantum mechanics, has a very surprising feature: the consequences of events at one place propagate to other places faster than light. This happens in a way that we cannot use for signalling. Nevertheless it is a gross violation of relativistic causality. Moreover the specific quantum phenomena that require such superluminal explanation have been largely realised in the laboratory ... especially by Aspect, Dalibard, and Roger, in Paris in 1982 (*Phys. Rev. Lett.* **49,** 1804 (1982)).
>
> For me then this is the real problem with quantum theory: the apparently essential conflict between any sharp formulation and fundamental relativity. That is to say, we have an apparent incompatibility, at the deepest level, between the two fundamental pillars of contemporary theory ... It may be that a real synthesis of quantum and relativity theories requires not just technical developments but radical conceptual renewal.

Popper has also considered the possibility that special relativity might have to be replaced and has suggested that its place might be taken by Lorentz's theory. Here is what he wrote[14]:

> It is only now, in the light of the new experiments stemming from Bell's work, that the suggestion of replacing Einstein's interpretation by Lorentz's can be made. If there is action at a distance, then there is something like absolute space. It should be remembered that nothing like a crucial experiment between Lorentz's theory and special relativity theory has ever been conceived before. If we now have theoretical reasons from quantum theory for introducing absolute simultaneity, then we would have to go back to Lorentz's interpretation.

In spite of the fact that Popper seems to favor Lorentz's theory as a possible replacement for Einstein's, there is also the possibility that a theory based on the new postulate that has been studied by

Moon, Spencer, and their colleagues might form the basis of a completely different theory; that is why it is important to discuss the whole subject thoroughly and objectively.

It should also be pointed out that there seems to be some disagreement among scientists about the significance of influences or signals that appear to travel faster than light, and whether they conflict with special relativity. For example, Rae[15] writes as follows:

> If these influences cannot be used to transmit information, then they need not be subject to the theory of relativity which requires that no *signal* can be transmitted faster than light. We are dealing with a *correlation* between two sets of events which does not travel in one direction or the other. [Italics in the original.]

There also seem to be wide differences of opinion about the significance of Bell's contributions. For example, Stapp[16] has stated that "Bell's theorem is the most profound discovery of science," while Vigier[17] has written, referring to the theoretical controversy and subsequent experiments that started with Bell's paper,[12] "Its experimental issue might prove as important for the future evolution of physics as the negative result of the Michelson-Morley experiment." By contrast, Mayants[18,19] claims that Bell's theorem is wrong, and writes[19] that "One may infer from Bell's theorem that an unrealistic supposition leads to a result incompatible with reality, a trivial conclusion." However, the esteem in which Bell's work is held by many scientists may be judged by the fact that the journal *Foundations of Physics* dedicated six successive monthly issues (October 1990 to March 1991) to Bell to mark his sixtieth birthday; his sudden death on 1 October 1990, just as those journals were appearing, is a great loss to science. The esteem in which Bell is held does not, of course, prove that he is correct. What is needed is open-minded discussion of his ideas as well.

While the problems of quantum theory have been debated at length for many years, the problems of special relativity have not received as much attention in the recent literature. Although the special theory has been subjected to a considerable amount of criticism for many years, and although the criticism has not all been satisfactorily answered, the scientific community as a whole has continued to believe strongly in the validity of the theory and does not appear to wish to resolve the problems that have been raised by critics of the theory.

For example, Dingle's criticisms[20] of the special theory have thrown the scientific community into a great deal of confusion, from which it has not yet emerged. Not only have different defenders of the theory given different and incompatible answers to Dingle's criticisms, as I have documented elsewhere,[21] they have also made different assessments of the cogency of his arguments. Some scientists have dismissed his arguments as being trivial, while others of comparable eminence have found themselves unable to understand them. Consider, for example, the following comments by two extremely eminent scientists (both Nobel laureates) on the same Dingle argument, namely, his claim that Einstein's original paper is self-contradictory because it can be used to show either that a moving clock runs slow, relative to a stationary clock, or that it runs fast.[22]

Born's published reply[23] included the following statement:

> Dingle's objections are just a matter of superficial formulation and confusion. The simple fact that all relations between space co-ordinates and time expressed by the Lorentz transformation can be represented geometrically by Minkowski diagrams should suffice to show that there can be no logical contradiction in the theory.

By contrast, here is part of what Professor P.M.S. Blackett (who was later President of the Royal Society) wrote to Dingle[24]:

> I am afraid I cannot help. I am naturally interested in relativity, particularly as I have taught the special theory for many years to our first-year students, and of course anybody dealing with cosmic rays or high energy particles uses it every day; one could not think accurately about any of these phenomena without using it. ... However, on the subtle point you are interested in I am afraid I have no contribution to make. I have often consulted my theoretical colleagues on this question, and can find none of them who has serious doubts about the ordinary formulation. I confess I cannot completely follow the details of yours and Max Born's argument, or rather, I have not had the time or inclination to do so.

When such an eminent scientist as Blackett, *who taught the special theory for many years,* confesses to being either unable or disinclined to follow the details of a criticism of the theory, it seems reasonable to suggest that the criticism is not trivial, and that there may indeed be a problem with the theory.

Few scientists appear concerned about the way in which Professor Dingle's criticisms have been treated by the scientific community, or about the way in which mainstream scientific journals have inhibited discussion of the subject. (A considerable amount of evidence on that inhibition of discussion can be found in Dingle's book.[20]) As a more recent example of the attitude of an influential member of the scientific community, here is what John Maddox, editor of *Nature*, has recently written about special relativity[25]:

> Given the zeal with which those who disbelieve the special theory of relativity go about their business, it is a comfort that experimental tests of relativity are also flourishing.

Maddox here appears to be implying that it is desirable for special relativity to be verified by experiment and undesirable that "those who disbelieve the special theory of relativity" should prevail. This impression is strengthened by the last sentence of the same article, which reads:

> Much has been done already, but further efforts are needed, and not merely to keep the army of sceptics of the special theory at bay.

It seems to me to be highly inappropriate for the Editor of a leading scientific journal to suggest or to imply that certain outcomes of scientific experiments are desirable for their own sake. One might sympathize with a suggestion that some particular scientific discovery (such as cold fusion) would be desirable from the point of view of human welfare, but not with a suggestion that it would be desirable for the universe to behave in such a way that all experimental results would satisfy the predictions of a particular theory. If the purpose of pure science is the discovery of scientific knowledge for its own sake, then the practitioners of such science should adopt a value-neutral attitude to possible outcomes of scientific experiments. The quest for truth is unlikely to be successful if it is assumed in advance that some possible truths are preferable to others.

Another interesting and disturbing feature of Maddox's comments is his apparent failure or unwillingness to realize that experimental results are not the point at issue, at least for many critics of special relativity. If the theory did in fact contain a contradiction, as many of the critics believe, then it would be possible to predict, by valid rules of inference from the postulates of the theory, any experimental results whatever. That means that no experimental results whatever can refute a claim that the theory contains an internal contradiction.

A similar error of logic is made by Ziman when he writes[26]:

> Thus, for example, anyone who would now set out to refute relativity theory could not succeed by, say, demonstrating a logical fallacy in Einstein's original paper. It would also be necessary to explain the successful use of this theory in a vast range of scientific work, ranging from the engineering design of particle accelerators to the fundamental equations of quantum physics.

On the contrary, discovery of a logical contradiction would be fatal to the theory, for the reason mentioned above. The comments of both Maddox and Ziman simply suggest that even some eminent scientists are rather poorly versed in logic. Born also shows his unsatisfactory command of logic in the passage quoted above, when he claims that "The simple fact that all relations between space coordinates and time expressed by the Lorentz transformation can be represented geometrically by Minkowski diagrams should suffice to show that there can be no logical contradiction in the theory." The Lorentz transformation is contained in the special theory; the special theory is not contained in (or identical with) the Lorentz transformation. Therefore, no property of the transformation can guarantee an absence of logical contradiction in the special theory. Although Born was probably the most eminent scientist ever to try to refute Dingle's claim that the special theory is self-contradictory, even supporters of the theory have found his answer unsatisfactory. For example, here is what Marder writes about Born's reply[27]:

> In a sense, it was a pity that Born then took up the challenge, because a satisfactory reply to Dingle needed more time than Born wished to devote to the matter. His brief reply, in *Nature,* consisted largely of a "correction" to Dingle's question (hardly likely to produce the desired effect) and a partially explained space-time diagram.

Born, by contrast, was completely satisfied with his own reply. According to Dingle,[28] he refused even to read Dingle's subsequent reply, claiming that his own argument was irrefutable. In taking this attitude Born seems somewhat inconsistent with the following excerpt from his Nobel Lecture, which seemed to his son to be so profound and important that he kept it framed on his desk[29]:

> I believe that ideas such as absolute certitude, absolute exactness, final truth, *etc.* are figments of the imagination

> which should not be admissible in any field of science. On the other hand, any assertion of probability is either right or wrong from the standpoint of the theory on which it is based. This *loosening of thinking* seems to me to be the greatest blessing which modern science has given to us. For the belief in a single truth and in being the possessor thereof is the root cause of all evil in the world. [Italics in the original.]

It appears that the resolution of the problems of special relativity and quantum mechanics may involve some revolutionary changes in science. It is therefore interesting to read what Wilson[(30)] has written about scientific revolutions. Since Wilson was actively involved in another scientific revolution, in connection with continental drift, his ideas on the growth and decay of scientific theories are worthy of serious study. He described the features that scientific revolutions possess in common, in the following way:

> In each case the great event marking the revolution revealed contradictions and introduced problems which the old theory could not resolve, leading to a period of confusion. For the most part it was not experts but rebels, outsiders and interlopers from other fields of science who suggested the need for a revolution and produced the evidence supporting it. Thus most geologists continued to oppose the concept of continental drift long after Wegener (a meteorologist) had championed it, after physicists had provided key evidence supporting it and after many of the public had accepted it.
>
> So, far from welcoming the new ideas, the establishment, who had the most to lose, clung for as long as possible to the old, justifying their position by questioning the new data, discrediting those who advanced them, and trying to patch up the old theories. In these endeavours they frequently found themselves supporting quite illogical positions. Before each revolution was accepted, the state of the subject had become chaotic, but each solved many problems so that the ensuing periods were times of great scientific progress and material benefit.

Although, as suggested by Popper in the passage quoted earlier, a crucial experiment between Lorentz's theory and special relativity may not have been conceived before, Dingle[(31)] suggested twenty years ago an experiment that would distinguish between Einstein's postulate and Postulate III. It would be a good idea to perform the experiment, if it is feasible to do so.

If, as seems more and more likely, special relativity has to be replaced by another theory, the resulting scientific revolution would cause considerable turmoil for a while, but it would also lead to an enormously stimulating period of scientific research. Scientists should not shrink from grasping such an opportunity.

Résumé

Un mémoire speculatif de Faraday en 1846 a acquis une importance renouvelée vis-à-vis des études récentes de Moon, Spencer et leurs collégues, dont le postulat III apparait être an alternative acceptable du postulat sur la vitesse de la lumière donné par Einstein en 1905. Une discussion impartielle du sujet est très importante, parce que les théories de la relativité restreinte et de la méchanique quantique sont impliquées.

References

1. M. Faraday, *Philos. Mag.* 3, **28**, 345(1846).
2. P. Moon, D. E. Spencer, and E. E. Moon: Phys. Essays **2**,275 (1989).
3. *Ibid.*, 368.
4. *Ibid.* **3**, 421 (1990).
5. P. Moon, D.E. Spencer, and S.Y. Uma, *Phys. Essays* **4**, 249 (1991).
6. P. Moon and D.E. Spencer, *Philos. Sci.* **23**, 216 (1956).
7. A. Einstein: *Ann. Phys.* **17**, 891 (1905).
8. H. Samuel and H. Dingle: *A Threefold Cord: Philosophy, Science, Religion,* (Allen & Unwin, 1961), p. 65.
9. A. Grünbaum, *Philos. Sci.* **24**, 77 (1957).
10. P. Moon and D.E. Spencer, *Philos. Sci.* **25**, 287 (1958).
11. A. Einstein, B. Podolsky, and N. Rosen: *Phys. Rev.*, **47**, 777 (1935).
12. J.S. Bell: *Physics* **1**, 195 (1964). [Reprinted in Ref. 13, p. 14.]
13. *Idem. Speakable and Unspeakable in Quantum Mechanics,* (Cambridge University Press. 1987), p. 171.
14. K.R. Popper, *Quantum Theory and the Schism in Physics,* (Hutchinson, 1982) p. 30.
15. A.I.M. Rae, *Quantum physics: Illusion or Reality?,* (Cambridge University Press.1986), p. 46.
16. H.P. Stapp, *Nuovo Cimento,* **B 29,** 270 (1975).
17. J.P. Vigier, *Lett. Nuovo Cimento,* **24**, 258 (1979).
18. L. Mayants, *The Enigma of Probability and Physics,* (Reidel, 1984).
19. *Idem. Phys. Essays,* **4**, 178 (1991).
20. H. Dingle, *Science at the Crossroads,* (Martin Brian & O'Keeffe, 1972).
21. I. McCausland, *Wireless World* **89,** 63 (October 1983).
22. H. Dingle, *Nature* **195,** 985 (1962).
23. M. Born, *Nature* **197,** 1287 (1963).
24. H. Dingle, *Science at the Crossroads*, p. 96.
25. J. Maddox, *Nature* **319,** 533 (1986).

26. J. Ziman, *An Introduction to Science Studies: The Philosophical and Social Aspects of Science and Technology*, (Cambridge University Press, 1984), p. 49.

27. L. Marder, *Time and the Space-Traveller*, (Allen & Unwin, 1971), p. 144.

28. Dingle, *Science at the Crossroads*, p. 43.

29. M. Born, *My Life: Recollections of a Nobel Laureate*, (Charles Scribner's Sons, 1978), p. 298.

30. J.T. Wilson, *Nature* **265,** 196 (1977).

31. H. Dingle, *Science at the Crossroads*, p. 210.

Appendix IV

An Inconsistency in Special Relativity*

This note presents an argument based on Einstein's original paper on special relativity, showing that the theory requires that each of two clocks in uniform relative motion actually works slower than the other. This shows an internal inconsistency in the special theory.

Key words: special relativity

About thirty years ago Dingle claimed that Einstein's special theory of relativity was internally inconsistent. In 1972 he published a book[1] describing his attempts to convince the scientific community that his argument was sound. The basis of his argument was his claim that the special theory required that if there were two clocks in uniform motion relative to one another, each one would actually work more slowly than the other, and that is physically impossible. The scientific community did not accept Dingle's claim of an inconsistency, one of the arguments used against him being that the special theory only required each clock to appear to work more slowly than the other; for example, Stadlen[2] wrote "What Einstein really said was that each clock would appear to run slow to an observer moving with the other."

Although there have been many other published claims that Einstein's special theory is untenable, the scientific community as a

* Reprinted from *Physics Essays*, Vol. 3, No. 2, 1990, pp. 357-361, by permission.

whole has not accepted those claims. In the author's opinion, the matter has not been satisfactorily resolved, and the purpose of the present note is to show, as simply as possible and using arguments taken directly from Einstein's original paper on the subject, that the special theory of relativity does indeed require that if there are two clocks in uniform relative motion, each one actually works slower than the other.

In his original paper on special relativity, Einstein[3] stated that if two clocks were together and synchronized and if one of them went around a polygonal line and returned to the other clock, then the one that had gone around the polygonal line would on its return show an earlier time reading than the one that had stayed behind. Clearly, that result cannot be merely a matter of appearance or observation; an observer who had gone around the polygonal line with one clock and an observer who had stayed behind with the other clock would agree that the difference in readings actually occurred.

Einstein then went on to assume that the result proved for a polygonal line is also valid for a continuously curved line and thence concluded that a clock at the equator would work more slowly than an exactly similar clock at one of the poles. It is again clear that the predicted slowing in this case is an actual slowing and not merely an effect of observation; observers attached to both clocks would agree that there was a progressive retardation of the equatorial clock, as compared with the polar clock, as the process continued. The fact that Einstein later added a footnote excluding the case of a pendulum clock emphasizes the reality of the phenomenon. (Although that footnote has been attributed to Sommerfeld, the present writer has argued elsewhere[4] that it was, in fact, written by Einstein himself.)

Now, taking a step backwards in the argument by which Einstein deduced that the equatorial clock would work more slowly than the polar one, suppose that the "equatorial" clock travels at uniform linear velocity along the perimeter of a square having the "polar" clock at its center. If *X* denotes the clock at the center of the square, and *Y* denotes the clock traveling along the square path, then, by Einstein's argument, *Y* would work more slowly than *X*. By symmetry *Y* would obviously experience the same amount of lag while traveling along each of the four sides of the square; furthermore, there would be no discontinuities of *Y*'s reading at any comer of the square. Those facts could be verified by repeated radio transmissions from Y to X, transmitting Y's reading as Y passed each corner of the square; the differences between successive transmitted readings would be of equal magnitude, denoted by T_Y, and the differences between successive readings of X's clock when the transmissions were received would also be of equal magnitude, denoted by T_X. According to Einstein's theory, T_Y would be smaller than T_X; in other words, Y would run steadily slower than X while traveling along any one side of the square.

Now consider a third clock, denoted by Z, traveling through space at uniform velocity relative to X in such a way that for a small portion of its journey it travels beside Y along one side of the square. Since it must keep time with Y as it travels beside it, then Z must also work steadily slower than X, even though its velocity relative to X is perfectly uniform.

In other words, if Y runs slower than X, then Z runs slower than X. Now, using the principle of relativity, it can with equal validity be shown that X runs slower than Z. These results, both of which are deduced from Einstein's original presentation of the theory, are inconsistent with one another.

In other words, if Einstein's prediction that a clock at the equator would work more slowly than a clock at one of the poles is valid, then each of two clocks in uniform relative motion must work more slowly than the other. This supports earlier claims that the special theory of relativity contains an internal inconsistency.

Received on 26 June 1989.

Résumé

Cette lettre présente une discussion fondée sur l'article original d'Einstein sur la relativité spéciale, montrant que la théorie exige que chacun de deux horloges en mouvement relatif uniforme marche plus lentement que l'âutre. Ceci montre une incohérence intrinsèque à la théorie.

References

1. H. Dingle, *Science at the Crossroads* (Martin Brian & O'Keeffe, London, 1972).
2. G. Stadlen, *The Listener* **88**, 411 (1972).
3. A. Einstein, "On The Electrodynamics of Moving Bodies," in H. A. Lorentz, A. Einstein, H. Minkowski, and H. Weyl, *The Principle of Relativity* (Methuen, London, 1923),
4. I. McCausland, *Br. J. Philos. Sci.* **35**, 60 (1984).

Appendix V

Anomalies in the History of Relativity*

In November 1919 it was announced to the world that observations of a solar eclipse that occurred in May 1919 supported Albert Einstein's general theory of relativity. That announcement was one of the most influential events of 20th-century science, since Einstein's instant rise to enormous fame arose directly from it. In spite of the confidence with which the announcement was made, however, it was later realized that the accuracy of the observations was insufficient to constitute a reliable confirmation of the phenomenon that was predicted. Furthermore, another of the formulas published in the general theory, for the variation in the perihelion of the planet Mercury, had already been derived by another scientist several years earlier using another method. In spite of the fact that the experimental evidence for relativity seems to have been very flimsy in 1919, Einstein's enormous fame has remained intact and his theory has ever since been held to be one of the highest achievements of human thought. The resulting deification of Einstein has had some unfortunate effects: critics of his theory are often dismissed as cranks, and the search for better theories has been inhibited. It is suggested that the announcement of the eclipse observations in 1919 was not a triumph of science as it is often portrayed, but rather an obstacle to objective consideration of alternatives.

Keywords: Eddington – Einstein – general relativity – gravitation – light deflection – solar eclipses

* Reprinted from "Anomalies in the History of Relativity," *Journal of Scientific Exploration*, Vol. 13, No. 2, 1999, pp. 271-290, published by the Society for Scientific Exploration, scientificexploration.org.

Introduction

The generally accepted version of the history of relativity has been dominated for about eighty years by the famous announcement in November 1919 that the results of observations at a solar eclipse in May 1919 confirmed Albert Einstein's prediction from his general theory of relativity that starlight would be bent by a precise amount by the gravitational field of the Sun. The announcement caused such an enormous sensation, both among scientists and the general public, that Einstein and his theory became world famous, literally overnight. In spite of the confident announcement, however, it was later realized that the observations were not particularly accurate and that the results did not in fact provide decisive support for the theory. An interesting commentary on the situation is found in the first two paragraphs of an article by the Editor of *Nature* (Maddox, 1995), which are as follows:

> Historically, the most dramatic proof that Einstein was on the right track with his general theory of relativity came in 1919, when Eddington and his associate Crommelin took themselves off to the western Pacific and the track of the total eclipse of the Sun in that year. They were bent on measuring the predicted deflection of light from distant stars lying near the limb of the Sun. As the world knows, the prediction was confirmed within the limits of the accuracy of the observations, to general applause. Einstein became a celebrity overnight.
>
> What is not so well remembered is that the measurements in 1919 were not particularly accurate. The standard error of the measurements was roughly 30 percent, while the displacement of the images of different stars in the field near the eclipsed Sun ranged from half to twice the deflection Einstein had predicted. And later eclipse measurements have not been much better, largely because of the difficulties of making accurate measurements from the *ad hoc* observatories that have to be established.

That quotation shows the two conflicting sides of the problem: Einstein's enormous and enduring fame resulted directly from the announcement of the eclipse results, although the results were not particularly accurate. Another interesting description of the problem is given by Sciama (1967):

> The first attempt to test Einstein's prediction was made by British expeditions led by Sir Arthur Eddington just at the end of the Great War. At that time it was considered very

> striking that a group of British astronomers should attempt to test a theory proposed by a German. The results of the test were claimed as a triumph for Einstein, and his world-wide fame dates from this event. Ironically enough, attempts made at later eclipses suggest that Eddington underestimated the uncertainties inherent in such a difficult observation, and even today Einstein's prediction has not been tested with the precision one would wish.

The purpose of the present paper is to present more details of the history of the period in question and to attempt to assess the significance of the announcement of the 1919 eclipse observations.

The General Theory of Relativity

The general theory of relativity (Einstein, 1916) gave a formula that correctly matched the anomalous advance of the perihelion of the planet Mercury, which had already been observed by astronomers such as Le Verrier (Roseveare, 1982). The general theory also made two further predictions: that light would be bent by a gravitational field, and that spectral lines of light from stars (including the Sun) would appear displaced towards the red end of the spectrum because of the strong gravitational fields.

The predicted effects of gravitation on light were so small that they caused no measurable phenomena on the earth itself, but the theory predicted that light reaching the earth from a star would be bent by the gravitational field of the Sun if the light passed very close to the Sun. Since a star whose light passed very close to the Sun would usually be impossible to see, the deflection of starlight by the Sun is very difficult to observe. However, such deflections might be observable during a total eclipse of the Sun, and in March 1917 the Astronomer Royal of Britain, Sir Frank Watson Dyson, suggested that the total eclipse of the Sun that was to take place on May 29, 1919, would present an excellent opportunity to test this prediction of the general theory (Dyson, 1917). Although 1917 was at a dark period of the first world war, Dyson undertook the organization of the eclipse expeditions in the hope that the war would be over in time for the expeditions to take place. Even though the war did in fact end in time for the expeditions, Dyson did not go to observe the eclipse himself. Instead, Professor Arthur Stanley Eddington became the leader of the eclipse expeditions, and the subsequent rise of the general theory was strongly influenced by Eddington's work. (As a minor comment on the above quotation from

Sciama, it should be pointed out that Eddington was knighted in 1930, so he was not Sir Arthur when he went on the eclipse expeditions.)

Although it appears that Dyson was originally somewhat skeptical about the possibility that Einstein's general theory would be confirmed (Eddington, 1939-1941), Eddington was a strong believer in the theory and was confident that it would be confirmed. Chandrasekhar (1987) has described Eddington's influence on Dyson by writing:

> Eddington's enthusiasm for the general theory of relativity must have succeeded in ensnaring into its magic his close friend and associate Sir Frank Dyson, the Astronomer Royal; for together, they were soon planning expeditions to observe the solar eclipse of 29 May 1919 if, to quote Jeans, "the state of civilization should permit when the time came." Eddington has described his own part in the planning and in the successful outcome of the expeditions "as the most exciting event, I recall, in my connection with astronomy."

Chandrasekhar went on to say that Eddington was so confident of the correctness of the theory that, if left to himself, he would not even have planned to go on the eclipse expedition. Chandrasekhar related how Eddington, a pacifist, was deferred from military service in the first world war "with the express stipulation that if the war should end by May 1919, then Eddington should undertake to lead an expedition for the purpose of verifying Einstein's prediction!"

The Eclipse Expeditions and their Observations

Two expeditions went from Britain to observe the total eclipse. Professor Eddington and Mr. E. T. Cottingham went to the island of Principe, in the Gulf of Guinea, West Africa, while Dr. A. C. D. Crommelin and Mr. C. Davidson went to Sobral in Brazil. Neither expedition went to the western Pacific as reported by Maddox (1995). Photographic plates of the stars close to the Sun, obtained during the eclipse, were compared with plates taken of the same portion of the sky at a different time of year when the Sun was elsewhere.

It should be noted that the deflection predicted by Einstein (1916) was very small, being about 1.7 seconds of arc for a ray of light at the edge of the Sun, and inversely proportional to the radial distance from the center of the Sun. Since no stars for which measurements were made during the 1919 eclipse were within two solar radii of the center, the largest deflection that would have occurred according to the theory would have been less than 0.8 seconds of arc, which, for the 343-cm focal length of the telescope used by Eddington, would have corresponded to about 0.01 mm on the photograph. In addition, there were other technical difficulties because of having to transport telescopes to remote locations. For example, light from the stars was reflected by a mirror into the telescope, so that the mirror could be rotated to compensate for the rotation of the earth during a time exposure, instead of rotating the telescope, which was not feasible under the conditions of the eclipse expeditions.

Possible sources of error in observations of this type at solar eclipses have been described by various authors such as von Klüber (1960) and Bertotti, Brill and Krotkov (1962). Some of these are:

1. Refraction of light in the Sun's corona and/or in the earth's atmosphere.
2. Distortions in the optical system caused by temperature changes during the eclipse.
3. Changes of scale between the eclipse plates and the comparison plates.
4. Distortions in the photographic emulsion while drying.
5. Errors in measurement of the images on the plates.

Although refraction in the Sun's corona was probably not significant, refraction in the earth's atmosphere may have been. For example, here is a recent estimate of the effect of the atmosphere on astronomical observations (MacRobert, 1995):

> Viewed at high power from the bottom of our ocean of air, a star is a living thing. It jumps, quivers, and ripples tirelessly, or swells into a ball of steady fuzz. Rare is the night (at most sites) when any telescope, no matter how large its aperture or perfect its optics, can resolve details finer than 1 arc second. More typical at ordinary locations is 2- or 3-arc-second seeing, or worse.

Although that article was in the "Backyard Astronomy" section of the journal, it seems reasonable to suggest that the conditions at Sobral and Principe in 1919 were probably not as good as the conditions obtainable in an American backyard observatory in 1995. The Sobral telescopes were set up on a race-course, and the set-up is shown in photographs in various publications such as Eddington (1920). The author is not aware of any published photographs of the Principe set-up, but it may be noted that the baggage had to be transported about a kilometer through the wood by native carriers (Dyson *et al.*, 1920), so the set-up was unlikely to have been any better than at Sobral. In addition to the ordinary everyday effects of the "ocean of air," the darkening of the Sun during the eclipse may itself have worsened the refractive effects.

Distortion in the optical system may have been caused by the effect of the heating of the mirror by the Sun, and this was noted as a problem at Sobral (Dyson *et al.*, 1920).

In considering changes of scale between the eclipse plates and the comparison plates, von Klüber (1960) has pointed out that, for a telescope with a focal length of 343 cm, a change in the effective focal length or focal setting of only 0.1 mm, between the eclipse plates and the comparison plates (which are taken months apart), can cause scale errors of the same order as the Einstein effect for stars beyond about 8 solar radii from the center of the Sun.

Distortion of the emulsion while drying was probably not very significant. However, measurements of the actual deflections on the plates may have introduced some errors in view of the extremely small displacements involved, especially when the x and y components of each displacement have to be measured.

In addition to all the above potential problems, the weather was cloudy at Principe at the time of the eclipse, and the photographs obtained were not as good as expected. In one of the earliest published references to the findings of the eclipse expeditions, the Astronomer Royal, Sir Frank Dyson, reported as follows at a meeting of the Royal Astronomical Society on July 11, 1919 (Fowler, 1919), before Eddington had returned from the expedition:

> I had a letter from Prof. Eddington two days ago. He is hoping to get good enough measures to determine the displacement definitely, but he obviously is greatly disappointed. He secured 16 photographs, but only for the last six was the sky clear enough to show any stars and on them he only got three, four, or five images; and, as the sky was generally only clear on one part of the plate at a time, the stars secured on the plates are badly distributed. From his best plate, however, he has some evidence of deflection in the Einstein sense, but the plate errors have yet to be fully determined.

It is interesting to compare that reference to Eddington's own later description of his measurement of that same plate. He describes how, although the weather was unfavorable because of cloud, sixteen photographs were taken. Of the twelve that were examined at Principe, most were useless, but "one plate was found showing fairly good images of five stars, which were suitable for a determination." (Eddington, 1920, p. 115) His description of this plate (p. 116) is as follows:

> The results from this plate gave a definite displacement, in good accordance with Einstein's theory and disagreeing with the Newtonian prediction. Although the material was very meagre compared with what had been hoped for, the writer (who it must be admitted was not altogether unbiassed) believed it convincing.

Eddington's biographer, Professor A. Vibert Douglas (1957), has also described his measurement of that plate, quoting from Eddington's Notebook. In the Notebook entry for June 3, after saying that he spent the whole day measuring, Eddington continued as follows:

> The cloudy weather upset my plans and I had to treat the measures in a different way from what I intended, consequently I have not been able to make any preliminary announcement of the result. But the one plate that I measured gave a result agreeing with Einstein.

Douglas's biography continued with the statement that "This was a moment which Eddington never forgot. On one occasion in later years he referred to it as the greatest moment of his life!"

It is interesting that the measurements on the same plate should have been viewed in such vastly different ways by Eddington at different stages of his study of the findings of the expedition: at first he was obviously greatly disappointed, then (although admittedly biased) he believed the evidence convincing in spite of the meager results, and later he saw it as the greatest moment of his life! Although he had at first suggested (in his letter to Dyson) that the stars on the plates were poorly distributed, he claimed in the published paper (Dyson *et al.*, 1920) that the five stars shown on the two plates from Principe "include all the most essential stars," and even suggested that the cloud may have been beneficial by preventing over-exposure of the brightest stars! In spite of that optimistic assessment, it appears from the published paper that the stars on the two usable plates obtained by the Principe expedition were indeed poorly distributed. Since the expected deflections, according to Einstein's theory, were radial, it seems obvious that a good test would include stars all around the Sun; but on the diagram showing the stars (Dyson *et al.*, 1920, p. 294), it is possible to draw a straight line through the center of the Sun in such a way that all the stars on both plates are well to one side of the line. In fact, it is possible to draw a straight line through the center of the Sun in such a way that, of the thirteen stars included in the combined records of both expeditions, only one (#2) is on the other side of the line. Furthermore, all five stars on one of the Principe plates (plate W, stars # 3, 4, 5, 6, 10), and four of the five on the other (plate X, the one measured by Eddington at Principe, stars # 3, 4, 5, 6, 11), are almost on a straight line. Since the measurements on the Principe plates were taken by placing a Principe plate film to film with an Oxford comparison plate so that the images of corresponding stars nearly coincided (which was possible because the Oxford plates were taken direct and the Principe plates by reflection in a mirror), the almost straight-line distribution of stars may have worsened the accuracy of fitting together the pairs of plates. (A diagram of the stars can also be found in Eddington (1920, p. 120).)

By contrast, the Sobral expedition had experienced fine weather; it had taken photographs using two telescopes, one similar to the Principe one and another having a different aperture and focal length. The photographs were also disappointing because the definition of the images had been spoiled. When the photographs taken by the telescope similar to the Principe one were measured "The measures pointed with all too good agreement to the 'half-deflection,' that is to say, the Newtonian value which is one-half the amount required by Einstein's theory." (Eddington, 1920, p. 117)

Eddington (1920) went on to give a reason for preferring the Principe results: the astronomers at Principe had taken "check-plates" of another portion of the sky to check whether there was any effect of the 50-degree difference in temperature between Principe in May and England in January, when the photographs were taken for comparison, whereas no such check-plates were taken by the Sobral expedition. However, the Sobral expedition had stayed in Brazil for a further two months after the eclipse (Eddington, 1920, p. 117) to photograph the same region of the sky before dawn, so it would appear that the large temperature difference would not apply to the comparison of the photographs in that case.

The published paper (Dyson *et al.*, 1920) states why it was not feasible for the Principe expedition to make a similar comparison:

> Unlike the Sobral expedition, we were not able to take comparison photographs of the eclipse field at Principe, because for us the eclipse occurred in the afternoon, and it would be many months before the field could be photographed in the same position in the sky before dawn. The check plates were therefore specially important for us.

In any event, it seems strange to state after the fact that the absence of check-plates of the Brazil expedition was a disadvantage. Either those check-plates were significant or they were not; if it was considered necessary to take them, that should have been decided *in advance* rather than being used *after the fact* to downgrade or reject a whole set of observations. A somewhat different opinion of the importance of the check-plates has been given by A. I. Miller (1996, pp. 88-89):

> Although the weather was far better at Sobral, Eddington insisted on emphasizing data from Principe. Clearly, a key part of this experiment is to use the same instruments at the same site for comparing data for the apparent and true star positions, but this was not done. The comparison photographs were taken several months later at Oxford.

There is also a puzzling discrepancy between the two accounts: although Eddington (1920, p. 117) states explicitly that "there were no check-plates taken at Sobral," the other account (Dyson *et al.*, 1920) states on p. 298, referring to the Sobral expedition, that "A few check plates of the field near Arcturus were taken, but have not been used."

Several authors have referred to the way in which Eddington emphasized some photographic plates rather than others. For example, Campbell (1923) wrote as follows:

> Professor Eddington was inclined to assign considerable weight to the African determination, but, as the few images on his small number of astrographic plates were not as good as those on the astrographic plates secured in Brazil, and the results from the latter were given almost negligible weight, the logic of the situation does not seem entirely clear.

As a minor addition to the history of the eclipse expeditions, it may also be noted that, at the same meeting at which Dyson reported Eddington's disappointment (Fowler, 1919), Campbell reported on the results obtained at an earlier eclipse on June 8, 1918, by saying "It is my own opinion that Dr. Curtis's results preclude the larger Einstein effect, but not the smaller amount expected according to the original Einstein hypothesis." (This refers to an earlier version of Einstein's prediction of the displacement, which was half the amount predicted by the general theory.)

Even after the measurements have been made, there is still the mathematical problem of extrapolating back to the edge of the Sun, since no stars measured were within two solar radii of the center of the Sun and small changes in the observed deflections can cause a much greater change in the computed deflection at the limb of the Sun. This problem has been discussed by von Klüber (1960) and Bertotti, Brill and Krotkov (1962); in particular, Figure 6 of von Klüber's paper shows extremely widely scattered results from various eclipse expeditions. A more recent eclipse expedition in 1973 has been described by Zirker (1995), who shows widely scattered measurements (Figure 9.4, p. 179) and states that the occurrence of a sandstorm limited the value of the results.

The problems of interpreting the results of the eclipse expeditions have been well summarized by Guggenheimer (1925), who made the following comment on the 1919 eclipse observations:

> Any reader, though far from an expert astronomer or physicist, who will study the description of the apparatus used in these observations and the large margin of error possible by reason of defects therein, will readily comprehend that, in view of the required delicacy of measurement of the things observed and of the error allowances both for apparatus defects and other possible physical causes of the observed phenomena, the greatest caution in the analysis of the results is necessary. Some astronomers deny that the photographs of the eclipse observations, when compared with those taken of the same stars in the absence of the sun, show deflections approximating the amount or the direction predicted by Professor Einstein. An examination of the various tables of the deflections observed shows that many of them are far away from the quantities predicted. The quantity approximating the predicted one is obtained by averaging a selected few of the observations.

Announcement of the Eclipse Results

The results obtained by the British eclipse expeditions of May 1919 were announced at the famous joint meeting of the Royal Society and the Royal Astronomical Society held on November 6, 1919. In spite of the poor accuracy and the uncertainties surrounding the results, it was announced that the evidence was decisively in favor of the value of displacement that had been predicted by Einstein, and Sir Joseph Thomson, President of the Royal Society and Chair of the meeting, strongly endorsed the results (Thomson, 1919). The enthusiastic reception of the announcement at the meeting was matched by an enormously strong reaction by the general public and, according to one biography of Einstein (Clark, 1971), "Einstein awoke in Berlin on the morning of November 7, 1919, to find himself famous." Although Fölsing (1997, p. 448) has expressed disagreement with that particular interpretation of the event, there is little doubt that Einstein rose with spectacular rapidity to worldwide fame as a direct result of the announcement.

The atmosphere of excitement at the meeting has been described by Alfred North Whitehead (1926) in the following words:

> It was my good fortune to be present at the meeting of the Royal Society in London when the Astronomer Royal for England announced that the photographic plates of the famous eclipse, as measured by his colleagues in Greenwich Observatory, had verified the prediction of Einstein that rays of light are bent as they pass in the neighbourhood of the sun. The whole atmosphere of tense interest was exactly that of the Greek drama: we were the chorus commenting on the decree of destiny as disclosed in the development of a supreme incident. There was dramatic quality in the very staging: the traditional ceremonial, and in the background the picture of Newton to remind us that the greatest of scientific generalisations was now, after more than two centuries, to receive its first modification. Nor was the personal interest wanting: a great adventure in thought had at length come safe to shore.

The meeting has also been described in a very interesting way by Abraham Pais (1982), who identified the day of the joint meeting as "the day on which Einstein was canonized." Pais was obviously very pleased with his comparison of the meeting to a Congregation of Rites at which a candidate is considered for canonization in the Catholic Church, and compared various participants at the meeting to counterparts in the Congregation of Rites, using as his reference *The New Catholic Encyclopedia*. However, one of his comparisons was highly inappropriate, namely his comparison of Ludwik Silberstein to the *advocatus diaboli*, or Devil's advocate. Although Silberstein was critical of accepting that the results superseded Newton's theory, the circumstances in which his criticisms were expressed were totally different, in at least two important respects, from the circumstances in which the arguments of the Devil's advocate are expressed in a Congregation of Rites.

In the first place, it is imperative that the arguments of the Devil's advocate be heard *before* canonization is pronounced, and Pais's account gives the impression that Silberstein's criticisms had been heard before the results were endorsed by the President of the Royal Society. However, the generally accepted account of the meeting (Thomson, 1919) shows that Thomson had, to use Pais's words, "pronounced the canonization" before Silberstein had had a chance to speak.

In the second place, it is the responsibility of the Devil's advocate to ensure that canonization does not occur undeservedly. In order to fulfill that responsibility, he must be given full access to all the relevant information required to make the case for the opposition to canonization. This condition was not fulfilled at the meeting at which the eclipse results were announced, because it was not possible for members of the audience to be sufficiently well-informed about the results that were being announced to make informed criticism of them. The paper that was eventually published (Dyson *et al.*, 1920), which is 43 pages long, with copious tables of results and mathematical analysis, carries the notation "Received October 30,–Read November 6, 1919." Also, the issue of *Nature* dated October 30 carried an announcement of the joint meeting, showing that the meeting had been arranged before the paper had been received by the Royal Society. It seems very unlikely, therefore, that the paper had received any critical review by independent referees before presentation, or that its contents were available to the audience early enough to be thoroughly studied. Just after the astronomers had presented their results, Thomson rose to call for discussion, but before the discussion actually started he strongly endorsed the confirmation of Einstein's prediction by saying (Thomson, 1919):

> It is difficult for the audience to weigh fully the meaning of the figures that have been put before us, but the Astronomer Royal and Prof. Eddington have studied the material carefully, and they regard the evidence as decisively in favour of the larger value of the displacement. This is the most important result obtained in connection with the theory of gravitation since Newton's day, and it is fitting that it should be

> announced at a meeting of the Society so closely connected with him. ...If it is sustained that Einstein's reasoning holds good—and it has survived two very severe tests in connection with the perihelion of Mercury and the present eclipse—then it is the result of one of the highest achievements of human thought.

It was that speech that Pais interpreted as the pronouncement of the canonization of Einstein; the remark about "one of the highest achievements of human thought" has been widely quoted and has obviously contributed enormously to the veneration of Einstein and relativity. If we pursue Pais's comparison of the joint meeting and a Congregation of Rites, we find another unfortunate feature of the comparison, namely that canonization of a saint by the Pope is infallible and irrevocable, so that subsequent criticism of the process is futile.

It is unfortunate that Pais, by identifying Silberstein as the Devil's advocate, gave the impression that a critical assessment of the eclipse results had been voiced at the joint meeting, and that the criticisms had been answered. Let us now consider what Silberstein did say in his contribution to the discussion at that meeting (Thomson, 1919). Although he could not find fault in the eclipse observations, for the reasons given above, he did point out the fact that the third main prediction of the general theory, the red-shift of light in a strong gravitational field, had not been observed. He said:

> There is a deflection of the light rays, but it does not prove Einstein's theory; it cannot be logically deduced from his theory as a gravitational effect in the absence of the spectroscopic result. And, as far as we know from St. John's and Evershed's observations, the predicted shift of the spectrum lines, of the amount exceeding almost 100 times the probable error of the modern spectroscope (as Prof. Fowler has just told us), is not obtained. ...If the shift remains unproved as at present the whole theory collapses, and the phenomenon just observed by the astronomers remains a fact awaiting to be accounted for in a different way.

Eddington's reply to Silberstein was described by Earman and Glymour (1980) as being "mild and irrelevant." He said:

> When a result that has been forecasted is obtained, we naturally ask what part of the theory exactly does it confirm. In this case it is Einstein's *law* of gravitation. It is not necessarily his theory that is confirmed, with the underlying assumption that *ds* is a quantity measurable by clock and scale.

> There still remains the question what the intermediary quantity *ds* is, which must be tested by the Fraunhofer lines. [Italics in the original.]

On the question of whether the eclipse results confirmed Einstein's general *theory* or only his *law* of gravitation, the published paper seems to be a little ambiguous; on p. 292 of the published paper (Dyson *et al.*, 1920) the following two statements occur within the course of a single paragraph (emphasis added):

> The results of the observations here described appear to point quite definitely to the third alternative, and confirm Einstein's generalised relativity *theory*.
>
> But, whether or not changes are needed in other parts of the theory, it appears now to be established that Einstein's *law* of gravitation gives the true deviations from the Newtonian law both for the relatively slow-moving planet Mercury and for the fast-moving waves of light.

If there is to be a fair comparison between the critical assessment of a scientific paper and the critical assessment of a candidate for canonization, as performed by a Devil's advocate, all available facilities should be placed at the disposal of the assessor in each case. Although Silberstein did not have enough information about the eclipse observations at the meeting itself, he made the following criticisms of them at a meeting of the Royal Astronomical Society on 12 December, 1919 (Fowler, 1920):

> They [the eclipse observations] indicate the presence of other factors modifying the displacement of the stars. These displacements are not radial. The deviations from the radial direction are marked running up to 15° for star No. 6 and to 35° for No. 11. Five of the stars near the Sun's axis show deviations from the radial displacement in the same sense; two stars near the Sun's equatorial plane show angular displacements in the opposite sense. If we had not the prejudice of Einstein's theory we should not say that the figures strongly indicated a radial law of displacement.

It seems reasonable to suggest that the prestige of Dyson and Eddington led to the arranging of the joint meeting before the paper had even been received, and to the rapid acceptance of the paper when it was received. In this connection, it is interesting to note the comment by Wali (1984, p. 115), about the influence of prestige on meetings of the Royal Astronomical Society, that "Papers had to be submitted a week before, by the first Friday of the month. Papers submitted by people like Eddington, Jeans and Milne were always read, and they always came first."

Although it seems obvious that no independent critical assessment of the results was made before the joint meeting, with information still available today it is possible to make some assessment of the results. As various authors such as von Klüber (1960) and Bertotti, Brill and Krotkov (1962) have pointed out, other scientists have re-assessed the results of the 1919 and other eclipse results, with a certain amount of variation from the results originally published. For example, Sciama (1969) refers to a table of results of eclipse observations from 1919 to 1952, and comments as follows:

> It is hard to assess their significance, since other astronomers have derived different results from a re-discussion of the same material. Moreover, one might suspect that if the observers did not know what value they were "supposed" to obtain, their published results might vary over a greater range than they actually do; there are several cases in astronomy where knowing the "right" answer has led to observed results later shown to be beyond the power of the apparatus to detect.

Hermann Bondi, a strong supporter of relativity, makes the following comment (Bondi, 1960) on the subject of Einstein's prediction of the deflection of starlight by the Sun:

> Einstein's prediction can therefore be checked only on the rare occasions when, at the moment of an eclipse, bright stars happen to be near the direction to the sun. The effect is difficult to study even when circumstances are most favourable. The indications are much in favour of Einstein's theory of relativity, but it would be premature to call the results conclusive.

If it was premature in 1960 to call the results conclusive, it was much more so in 1919.

Another interesting assessment of the significance of the 1919 eclipse results has been given by Calder (1979, p. 103), who wrote:

> The eclipse results were a triumph. Newton's ideas about gravity had reigned unchallenged for more than two centuries, yet within four years of Einstein developing his theory it seemed confirmed, and Newton was dethroned. The deflection of light by gravity is, as I have stressed, central to Einstein's general relativity. But later measurements of the deflection of starlight at other eclipses gave a wide scatter of results. They departed from Einstein's prediction by any-

> thing up to sixty per cent. The difficulties of the observations were to blame, rather than any defect in the theory. While they did not allow any restoration of Newton they left, nevertheless, a little elbow room for alternative accounts of gravity. So, sixty years after the initial triumph, astronomers and relativists were decidedly cool about this way of checking up on Einstein.

With reasoning like that, the theory cannot lose: if the experiment supports the theory, the theory is confirmed; if the experiment does not support the theory, it is the difficulties of the observations that are to blame, not the theory.

From reading the many accounts of the way in which the results were obtained, it appears that the evidence for the bending of starlight by the sun was rather flimsy in 1919. Let us now consider another of the results of the general theory, namely its explanation of the movement of the perihelion of the planet Mercury.

The Perihelion of Mercury

As already mentioned, the general theory of relativity included a formula that correctly matched the variation in the perihelion of the planet Mercury. Pais (1982) has given the following interesting description of Einstein's excitement on discovering his explanation of that phenomenon:

> This discovery was, I believe, by far the strongest emotional experience in Einstein's scientific life, perhaps in all his life. Nature had spoken to him. He had to be right.

Polkinghorne (1998) quoted Pais's statement and made the following comment:

> It was a great triumph but, if the answer had not come out right, the aesthetic power of the equations of general relativity would have been quite unable in itself to save them from abandonment. It was indeed *nature* that had spoken. [Italics in the original.]

It should be pointed out that exactly the same formula as Einstein published in his general theory for the variation in the perihelion of Mercury (Einstein, 1916) had been derived by another scientist several years earlier using another method based on the assumption of a retarded gravitational potential (Gerber, 1898). Some of the reactions to Gerber's paper have been described by Fölsing (1997). In particular, Fölsing described the action of physicist Ernst

Gehrcke, who had written articles against relativity since 1911, as follows:

> As no one believed his "refutations," Gehrcke in 1916 opened a second front. He unearthed some studies by the Pomeranian schoolmaster Paul Gerber, who, about the turn of the century, had tried to explain the perihelion movement of Mercury by procedures which were entirely arbitrary and were rightly rejected by astronomers. By reprinting Gerber's paper along with his own comment in *Annalen* Gehrcke insinuated, first, that Einstein was a plagiarist and, second, that the problem of the perihelion precession had been solved before relativity theory.

According to Fölsing, Einstein did not publish a reply to Gehrcke's paper, but he did publish a brief note (Einstein, 1918) in reply to another paper by Gehrcke (1918); however, although in that paper Gehrcke had mentioned Gerber's formula, Einstein's reply, which Fölsing described as "concise and factual," did not mention Gerber. Einstein did, however, mention him in a later article (Einstein, 1920), in which, after agreeing that Gerber had given the correct formula for the perihelion motion of Mercury before he himself had done, Einstein continued as follows:

> The experts are not only in agreement that Gerber's derivation is wrong through and through, but the formula cannot be obtained as a consequence of the main assumption made by Gerber. Mr Gerber's work is therefore completely useless, an unsuccessful and erroneous theoretical attempt. I maintain that the theory of general relativity has provided the first real explanation of the perihelion motion of Mercury. I have not mentioned the work by Gerber originally, because I did not know it when I wrote my work on the perihelion motion of Mercury; even if I had been aware of it, I would not have had any reason to mention it.

In the same article, Einstein referred to criticism of the eclipse observations as follows:

> Mr Gehrcke in his lecture has made the masterfully executed English measurements of the deflection of light by the sun appear in a bad light by mentioning the three independent groups only one of which, because of aberration of the heliograph mirror, gave erroneous results. He has suppressed the fact that the English astronomers themselves in their official report considered the results as a brilliant confirmation of the theory of general relativity.

In view of the many subsequent criticisms of the eclipse results, it appears that the confirmation of the theory may not have been as brilliant as those who obtained the results believed. It also seems unfortunate that Einstein based his claim of Gerber's errors on an appeal to the authority of "the experts" instead of indicating what was wrong with his derivation.

Another dismissal of Gerber's work is mentioned by Fölsing in the following words:

> A sharper, crushing reproof came from the respected Munich astronomer Hugo Ritter von Seeliger, who pointed out that the Pomeranian schoolmaster had copied things which had long been known to every worker in the field, and that his so-called explanation was based on a crude mathematical mistake.

Unfortunately, Seeliger (1918) does not seems to have identified the crude mistake, and Roseveare (1982) has made the following comment on his reply: "Seeliger claimed that Gerber's calculation was based on an elementary mistake, though it is evident that it was Seeliger who was mistaken." Unfortunately, also, Fölsing does not say why Gerber's methods, as he put it, "were entirely arbitrary and were rightly rejected by astronomers."It is also unfortunate that Gerber, having died sometime between 1902 and 1917 (Roseveare, 1982), was not available to defend his work from Einstein and Seeliger.

Since Gerber's 1898 paper and Gehrcke's references to it are in German, British scientists may not even have been aware of them in 1919, so that Gerber's work was not taken into account in considering possible alternatives to Einstein's theory. Simply pointing out Gerber's earlier derivation of the formula in question is not, of course, accusing Einstein of plagiarism, since Gerber's derivation of the formula was different from Einstein's.

Although Roseveare (1982) gives a sympathetic description of Gerber's work, he states that his law was refuted, on two counts. The first count was the fact that Gerber's theory is a gravitational theory and says nothing outside that sphere; this problem is apparently avoided by general relativity because, according to Roseveare, "this theory already contains special relativity." That claim is debatable at least, and we shall argue below, in our discussion of special relativity, that special and general relativity are distinct theories; neither one contains the other. Roseveare continues his claim that Gerber's theory has been refuted by saying "The second count on which Gerber's law may be seen to be refuted is the deflection of light rays, though this was at times rather equivocal." In view of the above discussion of the accuracy of the light-deflection results, this reason does not appear to be decisive. Perhaps it might be interesting to make an objective reappraisal of Gerber's work.

The Special Theory of Relativity

Although the above discussion concentrates on the general theory of relativity, there are also some anomalies in the history of the special theory. When the special theory was first published (Einstein, 1905), it was taken to be similar to theories of others such as Lorentz, Poincaré and Larmor. For example, Whittaker (1953) mentions Einstein's special theory in passing in a chapter titled "The Relativity Theory of Poincaré and Lorentz," in the following words:

> In the autumn of the same year, in the same volume of the *Annalen der Physik* as his paper on the Brownian motion, Einstein published a paper which set forth the relativity theory of Poincaré and Lorentz with some amplifications, and which attracted much attention.

Although the special theory did not have a very large impact before the rapid rise of the general theory in 1919, the events of 1919 caused it to become famous also, even though it is a different theory. Here is what Cohen (1985, p. 405) wrote about the way the events of 1919 contributed to the rise of the special theory:

> In considering relativity, we must keep in mind that there are two different theories of relativity. ...But what brought the attention of the world to special relativity was the verification in 1919 of a prediction of the general theory, that starlight passing near the sun is bent by the sun's gravity. This verification, which occurred during a solar eclipse expedition, set into being the relativity craze that swept the world and overnight made Einstein a public figure.

It is time to realize that, even if the general theory were verified, that does not in itself verify the special theory. The two theories are different; they do not use the same set of axioms. The difference is shown by the following statement by Stachel:

> 'Theory of relativity' is an umbrella term, covering two quite distinct theories, usually called 'the special' and 'the general' theory. The exact nature of the relation between the two is still a subject of controversy, and both might well not bear the common appellation 'relativity' if not so closely associ-

ated with the work of one man, Albert Einstein, who developed the second in attempting to extend the first to include gravitation.

Discussion

Various authors have tried to explain why the results of the eclipse observations were announced as being decisive when they were not. For example, Earman and Glymour (1980) write that

> Eddington's overenthusiastic advocacy may perhaps be explained by his prior conviction that the theory was true and by his interest in saving something from the vast work of the Principe expedition. Dyson's position might be understood in part as the result of a reasonable evaluation of the evidence, and in part as the result of Eddington's advocacy. But one retains the suspicion that besides these reasons, there was, especially for Eddington, another: the hope that a British verification of Einstein's theory would force on British scientists a more open-minded and generous attitude towards their German colleagues.

The enormous public reaction to the scientific announcement appears to have been strongly influenced by the ending of the war with Germany, and it may be noted that another important historical event of 1919 was the signing of the Treaty of Versailles, over German protest, in June. Chandrasekhar (1987) has described how Rutherford attributed Einstein's great fame to Eddington and explained the public reaction in the following words:

> The war had just ended; and the complacency of the Victorian and the Edwardian times had been shattered. The people felt that all their values and all their ideals had lost their bearings. Now, suddenly, they learnt that an astronomical prediction by a German scientist had been confirmed by expeditions to Brazil and West Africa and, indeed, prepared for already during the war, by British astronomers. Astronomy had always appealed to public imagination; and an astronomical discovery, transcending worldly strife, struck a responsive chord. The meeting of the Royal Society, at which the results of the British expeditions were reported, was headlined in all the British papers: and the typhoon of publicity crossed the Atlantic. From that point on, the American press played Einstein to the maximum.

It may be argued that the inaccuracies in the 1919 eclipse observations do not matter, since the same effect has supposedly been verified by more accurate measurements such as, for example, the measurements on quasars described by Maddox (1995). This point of view is somewhat guardedly espoused by the last paragraph of the critical discussion of the 1919 eclipse expeditions by Earman and Glymour (1980):

> This curious sequence of reasons might be cause enough for despair on the part of those who see in science a model of objectivity and rationality. That mood should be lightened by the reflection that the theory in which Eddington placed his faith because he thought it beautiful and profound–and, possibly, because he thought that it would be best for the world if it were true–this theory, so far as we know, still holds the truth about space, time and gravity.

That is at least debatable. For example, Chandrasekhar (1990) made the following comments on a lecture in which P.A.M. Dirac had claimed that there was a long and impressive list of successes of Einstein's theory of gravitation:

> It does not seems to me that the successes of Einstein's theory are either long or impressive. It is true that his prediction of the different rates of clocks in locations of differing gravity, his prediction of the deflection of light when traversing a gravitational field and resulting time delay, his prediction regarding the precession of the perihelion of Mercury, and finally, the slowing down of a binary star in an eccentric orbit by virtue of the emission of gravitational radiation, have been confirmed quantitatively. But all these relate to the departures from Newtonian theory by a few parts in a million; and of no more than three or four parameters in a post-Newtonian expansion of Einstein's field equations. And so far, no predictions of general relativity in the limit of strong gravitational fields have received any confirmation; nor are they likely in the foreseeable future.

It is also reasonable to ask whether the rapid and strong entrenchment of the general theory that occurred as a result of the eclipse announcement may have led experimenters to try to obtain the "right" answers from their observations, as suggested in the above quotation from Sciama (1969). For example, Collins and Pinch (1993) have commented on the measurements of the redshift predicted by general relativity as follows:

> The experimental observations, conducted both before and after 1919, were even more inconclusive. Yet after the interpretation of the eclipse observations had come firmly down

> on the side of Einstein, scientists suddenly began to see confirmation of the red-shift prediction where before they had seen only confusion.

Another unfortunate result of the announcement of the success of the eclipse observations has been an enormous hero-worship of Albert Einstein; Pais's statement that he was canonized has now been outmatched by Miller (1996, p. 90), who states that he was deified. A result of this deification is that the greatest scorn of the scientific community is reserved for those who would try to criticize either of Einstein's theories of relativity or to suggest alternative theories, and many mainstream scientific journals reject papers critical of either theory without review. The attitude of most journals is well described by Davies (1980) in the first sentence of his article "Why pick on Einstein?":

> Most editors of scientific journals make special provision for coping with the huge influx of papers and letters, many bearing private addresses in California, purporting to disprove or improve Albert Einstein's monumental work on the theory of relativity.

The assessment of scientific theories, and criticisms of them, should be based on the merits of each case, rather than on the addresses from which they come.

Although Maddox (1995) describes more modern ways of measuring the deflection of radiation by a gravitational field, he appears to be willing to consider the possibility that general relativity may not be the last word; the last paragraph of his article is as follows:

> The crying need remains what it has been for the past two decades, that of marrying together general relativity and quantum mechanics. As things are, they are like chalk and cheese. It will be a great surprise if general relativity survives that marriage unchanged.

As pointed out by Earman and Glymour (1980), it was Eddington's belief that confirmation of Einstein's prediction had a beneficial effect on scientific relations between England and Germany, and it is fair to ask whether that beneficial effect was sufficient to justify the announcement of the results as being decisively in favour of Einstein's theory when they were not. It is also fair to ask, today, whether continued belief in the decisiveness of the results as announced in November 1919 is beneficial to science and scientific objectivity, or whether scientific progress would be improved by a more open acknowledgment of the inaccuracies and uncertainties in the observations. Even if it is argued that Einstein's general theory has been supported by subsequent experimental observations, that does not alter the fact that, at an extremely important scientific meeting which had enormously far-reaching consequences, the audience was misinformed by eminent scientists about the phenomenon that was the main theme of the meeting. That historical fact is not wiped out by any subsequent experimental results, whatever they may be.

Because of the euphoric veneration of Einstein and relativity in November 1919, the objectivity with which science is supposed to act has been compromised, and the search for better theories has been inhibited. Canonization, deification, and claims of personal communication from Nature, should have no place in science. If the findings of the eclipse expeditions had been announced as being inconclusive instead of decisive in 1919, general relativity would have had to compete with other possible theories, such as Gerber's, to explain certain astronomical observations, and a better theory might eventually have been found. In the author's opinion, the confident announcement of the decisive confirmation of Einstein's general theory in November 1919 was not a triumph of science, as it is often portrayed, but one of the most unfortunate events in the history of 20th-Century science.

Acknowledgment

The author is grateful to Dr. Paul Marmet, Department of Physics, University of Ottawa, for bringing to his attention the work of Paul Gerber.

References

Bertotti, B., Brill, D., and Krotkov, R. (1962). Experiments on Gravitation, in Witten, L. (Ed.) *Gravitation: an introduction to current research.* New York: John Wiley, pp. 1-48.

Bondi, H. (1960). *The Universe at Large.* New York: Anchor Books.

Calder, N. (1979). *Einstein's Universe.* Harmondsworth, Middlesex: Penguin.

Campbell, W.W. (1923). The Total eclipse of the Sun, September 21, 1922. *Publications of the Astronomical Society of the Pacific*, 35, 11.

Chandrasekhar, S. (1987). *Truth and Beauty: Aesthetics and Motivations in Science.* Chicago: University of Chicago Press.

Chandrasekhar, S. (1990). *Relativistic Astrophysics*, (Selected Papers: S. Chandrasekhar, Vol. 5). University of Chicago Press.

Clark, R.W. (1971). *Einstein: The Life and Times*. New York: World Publishing Company.

Cohen, I.B. (1985). *Revolution in Science*. Belknap Press of Harvard University Press.

Collins, H., and Pinch, T. (1993). *The Golem: What Everyone Should Know About Science*. Cambridge University Press.

Davies, P. (1980) Why pick on Einstein? *New Scientist*, 87, 463.

Douglas, A.V. (1957). *The Life of Arthur Stanley Eddington*. London: Nelson.

Dyson, F.W. (1917). On the opportunity afforded by the eclipse of 1919 May 29 of verifying Einstein's Theory of Gravitation. *Monthly Notices of the Royal Astronomical Society*, 77, 445.

Dyson, F.W., Eddington, A.S., and Davidson, C. A (1920). Determination of the deflection of light by the Sun's gravitational field, from observations made at the total eclipse of May 29, 1919. *Philosophical Transactions of the Royal Society of London*, Series A, 220, 291.

Earman, J., and Glymour, C. (1980). Relativity and eclipses: The British eclipse expeditions of 1919 and their predecessors. *Historical Studies in the Physical Sciences*, 11, 49.

Eddington, A.S. (1920). *Space, Time and Gravitation: An Outline of the General Relativity Theory*. Cambridge University Press.

Eddington, A.S. (1939-1941). Sir Frank Watson Dyson. *Obituary Notices of Fellows of the Royal Society*, 3, 159-172.

Einstein, A. (1905). On the Electrodynamics of Moving Bodies. In Lorentz *et al.* (1923), pp. 37-65.

Einstein, A. (1916). The Foundation of the General Theory of Relativity. In Lorentz *et al.* (1923), pp. 111-164.

Einstein, A. (1918). Bemerkung zu E. Gehrckes Notiz "Über den Äther." *Verhandlungen der Deutschen Physikalischen Gesellschaft*, 20, 261.

Einstein, A. (1920). My Answer to the Anti-Relativistic Company, Inc. *Berliner Tageblatt und Handels-Zeitung*, August 27. In G.E. Tauber (Ed.) (1979), *Albert Einstein's General Theory of Relativity*. New York: Crown Publishers, Inc.

Fölsing, A. (1997). *Albert Einstein: A Biography*. New York: Viking.

Fowler, A. (1919). (Chair of) Meeting of the Royal Astronomical Society, Friday, July 11, 1919. *The Observatory*, 42, 297.

Fowler, A. (1920). (Chair of) Meeting of the Royal Astronomical Society, Friday, December 12, 1919. *The Observatory*, 43, 33.

Gehrcke, E. (1918). Über den Äther. *Verhandlungen der Deutschen Physikalischen Gesellschaft*, 20, 165.

Gerber, P. (1898). Die raümliche und zeitliche Ausbreitung der Gravitation. *Zeitschrift für Mathematik und Physik*, 43, 93.

Guggenheimer, S. H. (1925). *The Einstein Theory: Explained and Analyzed*. New York: Macmillan, pp. 298-9.

Lorentz, H.A., Einstein, A., Minkowski, H., and Weyl, H. (1923). *The Principle of Relativity*. London: Methuen.

MacRobert, A.M. (1995). Beating the Seeing. *Sky & Telescope*, 89, 40.

Maddox, J. (1995). More precise solar-limb light-bending. *Nature*, 377, 11.

Miller, A.I. (1996). *Insights of Genius: Imagery and Creativity in Science and Art*. New York: Springer-Verlag.

Pais, A. (1982). *'Subtle is the Lord...'*. Oxford: Clarendon Press.

Polkinghorne, J. (1998). *Belief in God in an Age of Science*. New Haven and London: Yale University Press.

Roseveare, N.T. (1982). *Mercury's Perihelion: from Le Verrier to Einstein*. Oxford: Clarendon Press.

Sciama, D.W. (1967). In *Einstein: The Man and His Achievement*, Edited by G.J. Whitrow, pp. 39-40. London: British Broadcasting Corporation.

Sciama, D.W. (1969). *The Physical Foundations of General Relativity*. New York: Doubleday.

Seeliger, H. (1918). Bemerkung zu dem Aufsatze des Herrn Gehrcke "über den Äther." *Verhandlungen der Deutschen Physikalischen Gesellschaft*, 20, 262.

Stachel, J. (1995). History of Relativity. In L.M. Brown, A. Pais, and B. Pippard, (Eds.) *Twentieth Century Physics*, Volume I, Chapter 4, pp. 249-356. Bristol: Institute of Physics Publishing.

Thomson, J. (1919). [Chair of] Joint Eclipse Meeting of the Royal Society and the Royal Astronomical Society. *The Observatory*, 42, 389.

von Klüber, H. (1960). The determination of Einstein's light-deflection in the gravitational field of the Sun. *Vistas in Astronomy*, 3, 47.

Wali, K.C. (1984). *Chandra: A Biography of S. Chandrasekhar*. University of Chicago Press.

Whitehead, A.N. (1926). *Science and the Modern World*. Cambridge University Press.

Whittaker, E. (1953). *A History of the Theories of Aether and Electricity: Volume two, The Modern Theories 1900-1926*. London: Nelson.

Zirker, J.B. (1995) *Total Eclipses of the Sun*. Princeton: Princeton University Press.

Appendix VI

The Einstein Mystique*

Albert Einstein's scientific career is studied, with the purpose of trying to explain why he became such a universally famous and revered person. Various events of the past century are considered, and their effects on his scientific and personal reputation. Some of the events studied are: the publication of the special and general theories of relativity, the 1919 solar eclipse and the famous meeting at which the results of the eclipse observations were announced, and Einstein's visit to the United States in 1921. After his death, many biographies of Einstein were written, both before and after the availability of further information that became available about his personal life after the deaths of Helen Dukas and Otto Nathan; some of these are discussed, including the strange story of what happened to Einstein's brain after his death. Celebrations of the centenary of his birth, the centenary of special relativity, and the centenary of the solar eclipse are also discussed. In spite of all the information that is available, the reasons for Einstein's great and enduring fame remain mysterious.

Keywords: Einstein — relativity — history of science

Introduction

Why is Albert Einstein so famous? Many writers have wondered why he is one of the most celebrated people who ever lived, and the answer is certainly not obvious. For example, Finch (1970) expressed the following opinion about the

* Reprinted from "The Einstein Mystique," *Journal of Scientific Exploration*, Vol. 17, No. 4, 2003, pp. 715-730, published by the Society for Scientific Exploration, scientificexploration.org.

problems that future historians may have in trying to explain the Einstein phenomenon:

> They may find themselves wondering how it happened that an abstract scientist, whose work could be understood only by a handful of people, should nevertheless have become an idol of millions so that his name and face were known all over the globe.

The history of twentieth-century physics was dominated by Albert Einstein. In the more than thirty years since Finch wrote the above words, Einstein's prestige has increased and is still increasing, especially as we approach the hundredth anniversary of his "miraculous year" 1905, in which he published five important papers. The activities for the celebration of the anniversary have already begun, starting with a major Einstein Exhibition that opened in The American Museum of Natural History in New York in November 2002, and which is to move to the Hebrew University of Jerusalem in time for the hundredth anniversary in 2005.

The Special Theory of Relativity

One of the papers that Einstein published in 1905 presented what is now known as the Special Theory of Relativity. The immediate impact of this theory, at the time it was published, is difficult to judge from the viewpoint of the present time, but the theory certainly became more prominent as a result of the subsequent rise of the General Theory. As pointed out by Fölsing (1997, pp. 201-3), the eminent scientist Max Planck was largely responsible for the acceptance for publication of Einstein's first paper on special relativity, and he was also "the most important figure in establishing relativity theory after 1905." Even so, the theory seems to have been less renowned outside Europe, judging by the following assessment that appeared in an editorial article in *Scientific American* (Anonymous, 1921):

> The Special Theory, promulgated fifteen years ago, received its fair share of attention from mathematicians all over the world, and is doubtless as well known and as fully appreciated here as elsewhere. But it has never been elevated to a position of any great importance in mathematical theory, simply because of itself, in the absence of its extension to the general case, it deserves little importance. It is merely an interesting bit of abstract speculation.

The General Theory of Relativity

Despite the fact that Einstein published five papers in 1905, the growth of his reputation was slow at first, and even after the publication of the General Theory of Relativity in 1916 that theory did not immediately become widely known. What caused the spectacularly rapid growth of his fame was the announcement, at a famous Joint Meeting of the Royal Society and the Royal Astronomical Society held in London on 6 November 1919, that the observations of an eclipse of the sun in May 1919 decisively supported his General Theory. The announcement was made by two very eminent astronomers: Sir F. W. Dyson, the Astronomer Royal, and Professor A.S. Eddington, Plumian Professor of Astronomy at the University of Cambridge, and resulted in such a spectacular rise in his scientific reputation that it was described by Pais (1982) as the canonization of Albert Einstein.

As I have described in detail in an earlier paper (McCausland, 1999), the accuracy of the 1919 eclipse observations was in fact not sufficient for a conclusive verification of the General Theory. Although that fact is now fairly widely known, scientists still appear to be very reluctant to admit it. As an example of the confusion that this reluctance causes, consider the following two quotations from the same scientist. In *A Brief History of Time*, Hawking (1988, p. 32) wrote:

> This proof of a German theory by British scientists was hailed as a great act of reconciliation between the two countries after the war. It is ironic, therefore, that later examination of the photographs taken on that expedition showed the errors were as great as the effect they were trying to measure. Their measurement had been sheer luck, or a case of knowing the result they wanted to get, not an uncommon occurrence in science. The light deflection has, however, been accurately confirmed by a number of later observations.

Several years later, in the "Person of the Century" issue of *Time* magazine, Hawking (1999, p. 41) wrote, referring to General Relativity:

> It was confirmed in spectacular fashion in 1919, when a British expedition to West Africa observed a slight shift in the position of stars near the sun during an eclipse. Their light, as Einstein had predicted, was bent as it passed the sun. Here was direct evidence that space and time are

> warped, the greatest change in our perception of the arena in which we live since Euclid wrote his *Elements* about 300 B.C.

Hawking's words, in the first of the above two quotations, illustrate some of the important features of the rise of the Einstein mystique. One of the reasons for Einstein's rapid rise to enormous fame was the supposed confirmation of a German scientist's theory by British scientists. The timing is especially important: World War I, one of the bloodiest events of the 20th Century, had ended in November 1918. The Paris Peace Conference, which lasted for six months, was the major historical event of 1919 and culminated in the signing of the Treaty of Versailles, over German objections, in June. The eclipse of the sun took place on May 29, 1919, and the announcement of the results took place on November 6, 1919, almost exactly a year after the armistice of November 11, 1918.

Various writers have described how the timing of the events affected the way in which the announcement of the results was received by the public. One very interesting account was given by Chandrasekhar (1987), who was present at an informal after-dinner discussion in Cambridge at which both Eddington and Ernest Rutherford were present. Someone said to Rutherford "I do not see why Einstein is accorded a greater public acclaim than you. After all, you invented the nuclear model of the atom; and that model provides the basis for all of physical science today and it is even more universal in its applications than Newton's laws of gravitation. Also, Einstein's predictions refer to such minute departures from the Newtonian theory that I do not see what all the fuss is about."

Rutherford turned to Eddington and said, "You are responsible for Einstein's fame." He went on to mention the war and the confirmation of a German scientist's predictions by British astronomers, and continued:

> Astronomy had always appealed to public imagination; and an astronomical discovery, transcending worldly strife, struck a responsive chord. The meeting of the Royal Society, at which the results of the British expeditions were reported, was headlined in all the British papers: and the typhoon of publicity crossed the Atlantic. From that point on, the American press played Einstein to the maximum.

The claim by Hawking (1988), that the light deflection has been accurately confirmed by later observations, has no relevance to the enormous impact that the announcement had on both the scientific community and the general public in 1919. The simple fact remains that Einstein was propelled to world-wide fame by an announcement that the observations decisively supported Einstein's theory, when the measurements were in fact, in Hawking's own words "sheer luck, or a case of knowing the result they wanted to get."

Another reason for the enormous impact of the announcement was the apparent incomprehensibility of the theory, which made those who claimed to understand the theory appear to be extremely clever. Supposedly there were only twelve men in the world who could understand it. A similar phenomenon has been described somewhat satirically by Galbraith (1990) in the words of one of the characters in his novel *A Tenured Professor*:

> Never forget, dear boy, that academic distinction in economics is not to be had from giving a clear account of how the world works. Keynes knew that; had he made his *General Theory* completely comprehensible, it would have been ignored. Economists value most the colleague whom they most struggle to understand. The pride they feel in eventually succeeding leads to admiration for the man who set them so difficult a task. And anyone who cannot be understood at all will be especially admired. All will want to give the impression that they have penetrated his mystification. This accords him a standing above all others.

With a few obvious minor substitutions, this quotation might be taken as an expression of the attitude of many physicists towards Einstein's General Theory.

After the enormous boost to Einstein's reputation that was caused by the joint meeting, the mystique surrounding the theory was maintained at its high level for some time by many further meetings devoted to the subject. The next strong boost to his fame appears to have occurred as a result of his visit to the United States in 1921. Goldberg (1984, p. 309) has described the reaction of the American public to that visit:

> After the initial interest in late 1919 and early 1920, a brief period of calm ensued which lasted until just before his first visit to the United States in 1921. If the proportion of space in the public press allotted to Einstein's comings and goings can be used as a measure of American fascination with him, then Einstein's visit here (on a fund raising tour on behalf of Hebrew University) was followed by the public in the

> manner of the arrival of the Beatles in the mid-nineteen sixties.

Another account of that visit was given by Missner (1985):

> When Einstein came to the United States in 1921 as part of a Zionist delegation, the warm welcome American Jews gave the delegation, and Chaim Weizmann in particular, was mistakenly described by the American press as a hero's welcome for Einstein. This led to a complex series of interactions between the Yiddish and English language press that resulted in Einstein being considered a hero and a secular saint.

Missner referred to the growth of Einstein's fame in the period from November 1919 to August 1921 as follows:

> This was the crucial period for the development of Einstein's fame, as afterwards his reputation grew and fed on what had already been achieved. While there is no reason to believe that there was a single cause for Einstein's fame, the American press was *the* instrument that made Einstein into a celebrity. [Italics in the original.]

In 1922 Einstein was awarded a Nobel prize for 1921, but not for relativity; the prize was awarded for his contributions to theoretical physics, including another of the papers that he had published in 1905, on the photoelectric effect. After a few more years there was little further contribution on Einstein's part, but his prestige remained extremely high. As Pais (1994, p. 43) wrote:

> Einstein is the only scientist to be justly held equal to Newton. That comparison is based exclusively on what he did before 1925. In the remaining 30 years of his life he remained active in research but his fame would be undiminished, if not enhanced, had he gone fishing instead.

Publications About Einstein

During the period from 1919 to Einstein's death, many books and articles were published about him and about relativity. After his death, many more biographies were published, one of the best being by Ronald W. Clark (1971). In a review of Clark's book, McCrea (1973) wrote:

> No one has yet explained the phenomenon that was Einstein, but if anyone ever does this it must surely be upon the basis of Clark's masterly account. The work has been done at just the right time. There can be few relevant documents

> that have not by now come to light and Clark seems to have digested all there are! The people who knew Einstein at all well have by now said all they will ever say about him.

Clark's book was an extremely thorough treatment of Einstein's life and work, based on the information available at the time he wrote. However, it subsequently turned out, as is described below, that there were indeed many documents that Clark either had not seen or was not allowed to use.

Most of the books and articles written about Einstein during the first twenty-five years after his death were highly laudatory, and in many cases Einstein was made to appear as saintly in his private life as he had been made a scientific saint at the joint meeting. Even his rather unpleasant divorce tended to be written off as an unfortunate minor irregularity, for example by Ziman (1981, p. 9):

> We admire also the lifestyle, so simple and modest, so kind and friendly, so courteous and good humoured, so liberal, wise and humane. Newton, as we are now informed, was arrogant, vain, suspicious and quarrelsome. It is difficult to imagine the character defects that will be discerned in our Einstein, 300 years from now. The failure of his first marriage seems no more than a mild accident of fortune.

Jaki (1989) drew attention to the adulation of Einstein during that period, saying that most of his biographers "portray him as an exemplary character without guile and with no fault." Jaki continued by saying that only on occasion do there appear in Einstein biographies "lines revealing for their brevity," giving a more realistic picture, and mentioned as an example a reference by Pais (1982, p. 14) to a letter written by Einstein admitting that he had disgracefully failed twice in marriage. Pais's book can be taken to mark the transition between the two kinds of biography: the earlier ones that show Einstein in a very favorable light, and the later ones that describe him more objectively. The main events leading to the changes in the kind of biography were the deaths of the two people who had joint control of his literary estate: Helen Dukas, his longtime secretary, and Otto Nathan. It is significant that Pais (1982) ends his Preface "To the Reader" in the following words: "I have left the text of this Preface as it was written before the death of Helen Dukas on February 10, 1982."

After the deaths of Dukas and Nathan, more information about Einstein and his private life became available, and it was revealed how carefully they had guarded his reputation. For example, as described by Highfield and Carter (1993, p. 269), Dukas had even been able to prevent Einstein's daughter-in-law Frieda from publishing a book based on his letters to his first wife, letters that were in Frieda's own possession. Highfield and Carter also pointed out that it was a source of regret to Nathan that he had allowed to be published "the most illuminating letter of all, in which Einstein admitted that he had 'failed rather disgracefully' in his marriages." They also pointed out that, although Clark had seen some of the revealing correspondence between Einstein and his friend Michele Besso before writing his biography of Einstein, he was not permitted to use the information. They mentioned some of Clark's problems on p. 275:

> Clark had been warned by Hans Albert to expect trouble from Nathan and Dukas, and he got it. ... He was threatened with legal action against both himself and his sources, and was refused key copyright permissions for the UK edition of his book. Although Clark fiercely resisted, Nathan forced him partially to rewrite the work before its publication in London in 1973.

(I assume that McCrea's review (1973), mentioned above, was based on the edition of Clark's book that was published in London in 1973.)

The biographical books that have appeared since Dukas's death, such as those by Brian (1996), Fölsing (1997), and Overbye (2000), have taken into account the further information that then became available about Einstein and his private life. With the appearance of these biographies there seemed to be little more to write about, and one might have thought that the Einstein publishing industry would become less prolific. However, the eagerness of authors to write about Einstein and his theories remains undiminished. If they cannot write on Einstein's life story as a whole, they write books about subjects such as *An Equation that Changed the World* (Fritzsch, 1994), *Einstein's Miraculous Year* (Stachel, 1998), *$E=mc^2$* (Bodanis, 2000), and *The Curvature of Spacetime* (Fritzsch, 2002). Alice Calaprice has published two books of quotations from Einstein, *The Quotable Einstein* (Calaprice, 1996), *The Expanded Quotable Einstein* (Calaprice, 2000), and a book of Einstein's letters to and from children (Calaprice, 2002). Another book of quotations (Mayer & Holms, 1996) bears the modest title *Bite-Size Einstein: Quotations on Just About Everything from the Greatest Mind of the Twentieth Century*. Recently Jerome (2002) has written a book based on the FBI file on Einstein; the file itself, which is about 1500 pages, is available on the internet site of the FBI. Jerome's book contains much interesting information on various unpopular causes that Einstein supported, and on the attempts of J. Edgar Hoover and Senator Joseph McCarthy to discredit him.

In view of the attention that is being paid in recent books to the equation $E = mc^2$, it should be pointed out that, although it is now a very famous equation, it was not the reason for the spectacular rise in Einstein's fame in November 1919, since few people would have realized at that time that the announced bending of light that caused his instant fame had anything to do with nuclear energy.

Einstein's Brain

One of the strangest stories associated with Einstein is the account of the adventures of his brain after his death. Although Einstein had expressed a wish that his body be cremated and his ashes scattered at an undisclosed location, in order to avoid the creation of a shrine, his brain was removed and kept by the pathologist who performed the autopsy, and it has since been treated in a way that almost certainly would have horrified Einstein.

The facts of the treatment of the brain have been presented by Carolyn Abraham (2001). Although the book documents the facts very well, it is unfortunately written in a highly rhetorical style with an abundance of irrelevant detail. For example, Einstein is variously described as "the smartest man this planet has ever produced" (p. 199), and "the most ubiquitous personality on the planet" (pp. 245-6), whatever that means. Some of the profusion of detail is shown in the following reference to the brain: "It grew in the belly of a German woman in the summer of 1878. It was barely a speck beneath skirts and petticoats and flesh, drawing forest-scented air from the Swabian Alps northwest of the Danube." A few pages later we are told that "It looked like a lima bean by the fall of 1878, just over a centimetre long." After Einstein's birth in 1879 the brain "feasted on its first gulps of oxygen," as if it had not received any oxygen from the forest-scented air the previous summer. Presumably it was the author's sense of delicacy that restrained her from starting the story of the brain from the gleam that appeared in Hermann Einstein's eye on a certain momentous day in 1878.

According to Abraham, even Einstein's family members were unaware that the brain had not been cremated with the body until an item appeared in a newspaper saying that the brain had been removed for scientific study. Einstein's son Hans Albert telephoned the Princeton Hospital to complain. Dr. Harvey, the pathologist who removed the brain, said that Hans Albert had given permission for an autopsy and, in Harvey's mind, "standard autopsy procedure included removal of the brain and, in some cases, keeping it." He appears to have subsequently received permission to keep the brain after promising that "it would only be used for scientific study and that reports about it would appear only in scientific journals." That would have been a difficult promise to keep, and the subsequent story shows that it has not been kept. Soon after Einstein's death there was some controversy between various scientists about who would be entitled to carry out research on the brain, and some of this was recorded in newspapers. As a sample of the reaction to conflicting reports of these controversies, the following is an extract from a letter from Otto Nathan to Dr. Harvey, dated April 26, 1955, just over a week after Einstein's death (Abraham, p. 85):

> I am at a loss to understand the various discrepancies and feel completely insecure about the various undertakings and promises made, and all of us are in addition gravely disturbed at the utterly distasteful notoriety, which would have shocked the late Professor Einstein beyond words.

If Einstein would have been shocked beyond words at the events surrounding his brain during the week after his death, those events were as nothing compared with what has occurred since that time. For example, thirty years later, Gina Maranto (1985) ended an article on Einstein's Brain with the following paragraph:

> Yet what lingers is a sense that the whole affair has become a tawdry botch. Especially considering neuroscientists' inability to discover from a brain the reason for genius, one wishes heartily that Einstein's brain had gone the way of his body, scattered to the winds of the universe, just another blip in the realm of space-time.

The tawdriness described by Maranto also seems a minor detail compared to the subsequent public coverage of the story of the brain. One of the oddest parts of the story, in my opinion, is the published account of the very strange car trip with the brain from

Princeton to California in February 1997, as told by Michael Paterniti, first in an article in *Harper's Magazine* (Paterniti, 1997) and subsequently in the book *Driving Mr. Albert* (Paterniti, 2000). When I first read the magazine article I thought that far too much print had been devoted to the story, and I was astonished that it later became a book. Now Abraham's book carries the even more astonishing story of a Paramount Pictures plan to make a movie of the Paterniti book.

One of the unfortunate features of Paterniti (2000) is a remarkable misunderstanding of what it was that Eddington supposedly observed at the solar eclipse in 1919. It appears on page 81, in the following words:

> Thus comes one of the huge revelations: Matter and energy drive the bending of space-time. To verify it, and almost as a triumphant toss-off, he produces an equation predicting the exact shift of sunlight as it passes Mercury—approximately forty-three seconds of arc per century, says Einstein—the very equation that, to the astonishment of the world, Arthur Eddington confirms as correct after observing the 1919 eclipse in western Africa.

The quoted passage confuses two completely different phenomena associated with the General Theory: the anomalous advance of the perihelion of the planet Mercury is forty-three seconds of arc per century, and requires years of observations to measure; Eddington's observations were completed in about five minutes of total eclipse, and measured the shift of starlight as it passes the sun, not the shift of sunlight as it passes Mercury.

Eventually Dr. Harvey returned Einstein's brain to the Princeton Medical Center in or about 1998, and it is interesting to read the descriptions of both Abraham and Paterniti of their subsequent visits to Dr. Elliot Krauss, pathologist at the Center and custodian of the brain. Both described how Krauss dismissed a paper that had been published about the brain, comparing it to phrenology. When Paterniti visited, Krauss had only had the brain for two days, and took it from under his desk to show it to him. When Abraham visited in February 2000, Krauss asked her to step outside the office while he fetched the brain from a secret location (possibly under his desk) to show it to her.

It should also be mentioned that Einstein's eyes were removed by another doctor, and are kept in a safety-deposit box in a bank. In Paterniti's book they are in a bank in Philadelphia, in Abraham's they are in a bank in New Jersey; perhaps they may have been moved between the writing of the two books. It appears that Einstein's brain and eyes are being treated like relics of a saint instead of ordinary physical objects.

Personally, I am appalled at the indignity that has been done to the memory of Albert Einstein in the way that his brain has been treated, with so little contribution to knowledge as a result. I believe that Einstein would have felt that the treatment of his brain and his eyes, in the way described by Abraham in her book, was completely unacceptable, and I would urge those who have them in their possession to have them cremated and scattered as he wished to have done with his body.

Perhaps it might also be appropriate to express the hope that Stephen Hawking has stipulated very clearly what is to be done with *his* brain when he has no further use for it, since for obvious reasons it would be a very interesting brain to study.

One of the interesting features of Paterniti (2000, pp. 170-4) is the story of his visit to the Beverly Hills office of Roger Richman, "president of his own celebrity-licensing agency and the man who represents the beneficiaries of the estate of Albert Einstein, Hebrew University in Jerusalem." Richman described Einstein as "the most widely recognized human being that ever lived," and the scale of his Einstein operation may be judged from Paterniti's assessment:

> Richman won't reveal how much money he and Hebrew University make annually from Einstein, but he admits it's more than from any other client, totalling somewhere in the seven figures.

Abraham (2001, pp. 278-9) also mentioned the Richman agency and described how Otto Nathan, three years before his death, had arranged with the agency to license Einstein's image for the benefit of the Hebrew University of Jerusalem:

> The agent eventually told Nathan that Einstein was in such hot demand that the estate ought to have some control over the appropriate use of his image and receive some benefit for allowing it. For Nathan, as always, exercising some control was better than no control. He put Richman in touch with the Hebrew University. To this day, Einstein ranks as Richman's hottest property.

The Religious Significance of Relativity?

Another very interesting phenomenon that appears in connection with Einstein is the way in which scientists have tried to take over God from the religious communities, or to use God to sell their books about the scientific saint. An interesting example is Pais's description of the choice of the title of his first biography of Einstein (Pais, 1982): '*subtle is the Lord...*'. Pais (1997) has described how his publisher was reluctant to use that title because the book would then end up on religion shelves in bookstores, to which Pais replied "So what? That will mean selling even more books." Subsequently there has appeared the book *God's Equation*, by Amir D. Aczel (1999), in which there appears an equation (p. 218) described as "Einstein's field equation with the cosmological constant, which is our best estimate of God's Equation." Aczel ends his book with the following paragraph:

> Once each discipline is supported by developments in the others, we may begin to understand the ultimate laws of nature and to formulate our human estimate of God's Equation. When the final equation is constructed, we should be able to use it to solve the wonderful riddle of creation. And perhaps that's why God sent us here in the first place.

It seems very strange to me to suggest that God might have created us just so that we could spend our time trying to solve the problem of why he created us.

Another book that associates Einstein with religion is Corey S. Powell's *God in the Equation* (2002). The link with Einstein appears in the subtitle: *How Einstein Became the Prophet of the New Religious Era*. The hyperbole seems even more intense than in most other books on Einstein, a typical sample of which appears in the first two sentences of the description on the book jacket:

> We are living at a turning point in human spirituality–akin to when Jesus or Buddha or Mohammed was alive–and Einstein is its prophet. That is the audacious, provocative, and fascinating argument Corey Powell makes with dazzling eloquence in this extraordinary book.

The author makes somewhat extravagant claims about what he calls the Church of Einstein and the new faith of sci/religion that is associated with it. We are told (p. 246) that "The Church of Einstein is more authoritative and comprehensive than ever. Its spiritual power has eclipsed that of the old-time religions." Many scientists apparently worship in the Church of Einstein; for example we read on p. 245: "At the American Astronomical Society, as well, there is more sci/religious zeal than the superficial sights and sounds indicate. Despite their reluctance to invoke overtly theological language, in their actions the current priests of sci/religion remain firmly committed to the Church of Einstein."

The recent books about Einstein seem to be even more adulatory than the ones that appeared before his private life became public knowledge, and Einstein's admirers still appear to be strongly motivated to enhance his eminence. Judging from many recently published books, several publishers have decided that a photograph of Einstein on the cover and/or Einstein's name in the title or the subtitle is a sure way of selling a book; presumably it gives readers an impression that they must be very clever to be able to read a book that has some association with Einstein. Mentioning the deity in the title is an additional way of increasing sales.

Greatness and Celebrity

In any case, the big question is still: why is Einstein such a revered figure? Even if the astronomers had been justified in their claim at the famous Joint Meeting that the observations conclusively supported the general theory, it is still difficult to understand why the verification of such an abstruse scientific theory caught the imagination of the general public to such an enormous extent in 1919. A possible reason was given by Clive James (1993), who wrote that "Einstein achieved world fame, largely because of the popular though erroneous notion that his theories of relativity had some relevance to ordinary life." Another interesting clue to his fame is a remark by Abraham Pais (1994, p. 139): "To Einstein applies *par excellence* the whimsical yet profound definition of a celebrity: a person who is famous for being well-known." That statement forms an excellent link to Daniel J. Boorstin's discussion of the difference between greatness and celebrity that comprises Chapter 22 (From Hero to Celebrity) of his book *Hidden History* (Boorstin, 1987). Boorstin uses a very similar phrase to the one used by Pais: "*The celebrity is a person who is known for his well-knownness.*" [Italics in the original.] Fame or celebrity can be manufactured; greatness can not, and one mark of the difference between a hero and a celebrity is given as follows:

> The hero was distinguished by his achievement, the celebrity by his image or trademark. The hero created himself; the celebrity is created by the media. The hero was a big man; the celebrity is a big name.

It may be instructive to consider another famous person, whom Boorstin gives as a dramatic, tragic example of the distinction between greatness and celebrity: Charles A. Lindbergh. Like Einstein, Lindbergh became world-famous almost overnight, having performed one of the most spectacular feats of the Twentieth Century when he flew the Atlantic Ocean alone from New York to Paris. As Boorstin points out, "Except for the fact of his flight, Lindbergh was a commonplace person." Boorstin records that, when Lindbergh returned to New York, "the *New York Times* gave its first sixteen pages the next morning almost exclusively to news about him" and goes on to say:

> Lindbergh was by now the biggest human pseudo-event of modern times. His achievement, actually because it had been accomplished so neatly and with such spectacular simplicity, offered little spontaneous news. The biggest news about Lindbergh was that he was such big news. Pseudo-events multiplied in more than the usual geometric progression, for Lindbergh's well-knownness was so sudden and so overwhelming. It was easy to make stories about what a big celebrity he was, how this youth, unknown a few days before, was now a household word, how he was received by Presidents and Kings. There was little else one could say about him.

Boorstin goes on to point out how Lindbergh remained a celebrity mainly because of two subsequent events: his marriage to Anne Morrow, daughter of the American Ambassador to Mexico, and the kidnapping and murder of his infant son. However, his celebrity eventually diminished rapidly, as explained by Boorstin:

> Democratic faith was not satisfied that its hero be only a dauntless flier. He had to become a scientist, an outspoken citizen, and a leader of men. His celebrity status unfortunately had persuaded him to become a public spokesman. When Lindbergh gave in to these temptations, he offended. ... His pronouncements were dull, petulant, and vicious. He acquired a reputation as a pro-Nazi and a crude racist; he accepted a decoration from Hitler. Very soon the celebrity was being uncelebrated.

Galbraith (1990) had a satirical explanation for this phenomenon also: "In all these cases there was a lesson: find out who in any euphoric episode is the greatest hero, who is the most celebrated, and invest in his eventual fall."

Like Lindbergh, Einstein's celebrity status was maintained by subsequent events, the main ones being his flight from the Nazis and his immigration to America, his letter to President Roosevelt about the nuclear bomb, and the development and use of the bomb. As in Lindbergh's case, society was not satisfied that its hero be only a clever scientist; Einstein was for the rest of his life taken to be a kind of oracle whose opinion was valuable in all fields of life, not just in physics. That may have been an undesirable result of his fame also, because the opinions of scientists outside their specialties are not necessarily any better than those of ordinary people. A rather strong statement to that effect was made by Jacques Ellul (1964, p 435):

> We are forced to conclude that our scientists are incapable of any but the emptiest platitudes when they stray from their specialties. It makes one think back on the collection of mediocrities accumulated by Einstein when he spoke of God, the state, peace, and the meaning of life. It is clear that Einstein, extraordinary mathematical genius that he was, was no Pascal; he knew nothing of political or human reality, or, in fact, anything at all outside his mathematical reach. The banality of Einstein's remarks in matters outside his specialty is as astonishing as his genius within it. It seems as though the specialized application of all one's faculties in a particular area inhibits the consideration of things in general.

Hook (1987) also criticized Einstein's opinions in fields other than science. In his chapter "My Running Debate with Albert Einstein" he said "there is a tendency, especially in modern times, to give undue weight to the political judgments of great figures in art, literature, and science." He felt that the tendency was greater in the case of scientists, "primarily because their vocation seems to express the quintessential practice of rationality."

Why is Einstein so Famous?

It still seems difficult to understand why Einstein has held the attention of the world for so long. Was he just a celebrity, or was he really a great man? McCrea (1973), who was an admirer of Einstein and a strong defender of relativity against its critics, wrote: "I think he was certainly not in the class of Newton, and personally I question whether he was as great a scientist as Faraday or Maxwell or Rutherford or Bohr." Yet those scientists did not become especially famous for their accomplishments. As Missner (1985) pointed out:

> Einstein's achievements, as great as they were, cannot be used to explain why other scientists whose achievements were of at least similar magnitude, such as Niels Bohr and Werner Heisenberg, did not gain any public recognition at all.

It is interesting to compare McCrea's opinion with the following assessment by Cornelius Lanczos (1965):

> He discovered a host of other basic results in theoretical physics and it has been pointed out more than once that if somebody asked: "Who is the greatest modern physicist after Einstein?" the answer would be: Einstein again. And why? Because, although the theory of relativity in itself would have established his fame forever, had somebody else discovered relativity, his other discoveries would still make him the second greatest physicist of his time.

That is a remarkable statement; it immediately makes every other modern scientist, no matter how eminent, drop to third place or lower, behind Einstein and "Zweistein."

A reviewer (Anonymous, 1974) of another adulatory book by Lanczos made the following comment:

> No doubt we can easily forgive the author's over-anxiety about the status of his hero, but the constant jumping-up-and-down exhibited by recent writers on Einstein is too desperate by far and too noisy. They protest too much. Surely Einstein himself, it is safe to assume, is waiting serenely and quietly for everything to take its proper place.

What *is* the proper place for everything to take? Even Einstein did not expect his theory of relativity to last for ever; his own opinion was stated in *The Evolution of Physics* (Einstein and Infeld, 1938) as

> There are no eternal theories in science. It always happens that some of the facts predicted by a theory are disproved by experiment. Every theory has its period of gradual development and triumph, after which it may experience a rapid decline.

In a review of another book on Einstein, McCrea (1982) wrote: "Actually, two chapters seem still to be missing from all the books about Einstein. One of these would tell of his fantastic *luck* in his scientific work—using the word as in ordinary parlance." [Italics in the original.] He went on to say: "Einstein was one of the greatest physicists who ever lived, but even Einstein's scientific career depended for much of its success upon countless circumstances beyond his control."

Einstein was indeed lucky in at least two instances: first, it was the prestige of Max Planck that helped to establish his Special Theory, and second, it was the prestige of Dyson and Eddington that caused the eclipse measurements (which were themselves "sheer luck") to be accepted so readily at the joint meeting in 1919.

One of the main reasons for Einstein's continuing to be such a celebrity, nearly a century after the publication of his first paper on special relativity, is that the news media have kept his fame alive. For example, *Time* magazine awarded him the status of Person of the Century in December 1999. An amusing token of the incomprehensibility of his work may be noted in a photograph accompanying the article by Hawking (1999) in the "Person of the Century" issue. The photograph, which shows Einstein lecturing in Paris in 1922, is printed the wrong way round so that the equations on the blackboard run from right to left! Correctly printed copies of the same photograph can be found in Pais (1982, after p. 272) and Fölsing (1997, after page 242).

Einstein has even been voted Person of the Millennium. A Canadian newspaper, *The Globe and Mail*, published the results of a readership survey on January 2, 1999, with Einstein as the readers' choice as the person who had "the most impact in the past 1,000 years." The same newspaper published on January 1, 2000 the results of another survey in which readers voted for their person of the millennium; in this case Johann Gutenberg (who had been eighth in the earlier survey) was first and Einstein second.

It is reasonable to suggest that, of all the people who voted for Einstein as Person of the Century or the person who had "the most impact in the past 1,000 years," very few really understood what Einstein really accomplished and exactly what impact his accomplishments have on their daily lives. Public perception of the various choices for the person of the period in question is strongly based on what the media say about the candidates and on the persuasive powers of those who solicit support for them in the voting. As an example of what can happen when greatness is measured by voting, and an illustration of the difference between greatness and celebrity, consider a nationwide poll held by the British Broadcasting Corporation in 2002 to find the greatest Briton, the results of which were given in *The Times* of London (on-line edition, November 25, 2002). Winston Churchill came first. Isambard Kingdom Brunel came second, his candidacy having been helped by an organized campaign by students at Brunel University. Diana, Princess of Wales, came third, followed by Charles Darwin, William Shakespeare, Isaac Newton, Queen Elizabeth I, John Lennon, Horatio Nelson, and Oliver Cromwell, in that order.

Einstein's fame has also been kept alive by elaborate and widespread celebrations in 1979 of the hundredth anniversary of his birth. Several books were published in honor of the occasion, but they were not all of uniformly high quality. One frustrated reviewer (Hendry, 1982) made the following comment in a review of three commemorative books about Einstein that were published at about that time:

> Question: How do you induce a perfectly sane and intelligent person to talk absolute rubbish?
>
> Answer: Ask him or her to talk about Einstein.

No doubt there will be many more books and articles published in connection with the centennial of Einstein's Miraculous Year, and it will be interesting to see how the quality and quantity compare with those published for the centennial of his birth. Although it appears from the publicity for the current Einstein Exhibition that the darker side of Einstein's personal life is being portrayed, it still seems doubtful whether there is to be an objective appraisal of the observations at the 1919 solar eclipse. If there is to be a celebration of the hundredth anniversary of the 1919 eclipse and the Joint Meeting, as presumably there will be, it is to be hoped that an honest assessment of the eclipse observations will be clearly made known in whatever celebrations take place at that time.

Why is Albert Einstein so famous? A typical attitude of perplexity was well expressed by Rowe (1973) in the last paragraph of his interesting review of Clark's biography:

> I must confess I have not reached a wholly satisfactory answer to the problem of why Einstein was, is, and is likely to be for centuries to come, one of the immortals. If the reader tries to find a solution he will find the task more absorbing than any detective story.

I share Rowe's perplexity, but not his opinion of the longevity of Einstein's towering reputation. Although there are many features of Einstein's story that can be adduced to try to explain his enormous and continuing fame, that fame is so overwhelming that no explanation seems completely satisfactory. It is indeed an absorbing and fascinating story and, like Rowe, I commend the problem to the reader for further study.

References

Abraham, C. (2001). *Possessing Genius: The Bizarre Odyssey of Einstein's Brain*, Harmondsworth, Middlesex: Penguin, Viking.

Aczel, A. D. (1999). *God's Equation: Einstein, Relativity, and the Expanding Universe*. New York: Four Walls Eight Windows.

Anonymous (1921). "The Einstein Award," *Scientific American*, January 29, p. 82.

Anonymous (1974). "That man again," (Review of Lanczos (1974)), *The Economist*, Vol. 252, No 6836, August 31, p. 81.

Bodanis, D. (2000). *E=mc²: A Biography of the World's Most Famous Equation*. Toronto: Doubleday Canada.

Boorstin, D. J.(1987). *Hidden History*, New York: Harper and Row.

Brian, D. (1996). *Einstein: A Life*. New York: Wiley.

Calaprice, A. (1996). *The Quotable Einstein*, Princeton, NJ: Princeton University Press.

Calaprice, A. (2000). *The Expanded Quotable Einstein*, Princeton, NJ: Princeton University Press.

Calaprice, A. (2002). *Dear Professor Einstein: Albert Einstein's Letters to and From Children*, New York: Prometheus Books.

Chandrasekhar, S. (1987). *Truth and Beauty: Aesthetics and Motivations in Science*. Chicago: University of Chicago Press.

Clark, R. W. (1971). *Einstein: The Life and Times*. New York: World Publishing Co.

Einstein, A., and Infeld, L. (1938). *The Evolution of Physics*, Cambridge: Cambridge University Press.

Ellul, J. (1964). *The Technological Society*. New York: Vintage Books.

Finch, H. Le R. (1970). "Introduction," in Moszkowski, A. *Conversations With Einstein*, New York: Horizon Press.

Fölsing, A. (1997). *Albert Einstein: A Biography*. New York: Viking.

Fritzsch, H. (1994). *An Equation that Changed the World: Newton, Einstein, and the Theory of Relativity*, Chicago: University of Chicago Press.

Fritzsch, H. (2002). *The Curvature of Spacetime: Newton, Einstein, and Gravitation*, New York: Columbia University Press.

Galbraith, J. K. (1990). *A Tenured Professor*. Boston: Houghton Mifflin.

Goldberg, S. (1984). *Understanding Relativity: Origin and Impact of a Scientific Revolution*. Boston: Birkhäuser.

Hawking, S. (1988). *A Brief History of Time: From the big bang to black holes*. London: Bantam Press.

Hawking, S. (1999). "A Brief History of Relativity," *Time*, (Canadian Edition) December 31, p. 28.

Hendry, J. (1982). Review, in *British Journal for the History of Science*, July, pp. 200-201.

Highfield, R., and Carter, P. (1993). *The Private Lives of Albert Einstein*. London: Faber and Faber.

Hook, S. (1987). *Out of Step: An Unquiet Life in the 20th Century*. New York: Harper and Row.

Jaki, S. L. (1989). *God and the Cosmologists*. Edinburgh: Scottish Academic Press.

James, C. (1993). *Fame in the 20th Century*, London: BBC Books.

Jerome, F. (2002). *The Einstein File: J. Edgar Hoover's Secret War Against the World's Most Famous Scientist*, New York: St. Martin's Press.

Lanczos, C. (1965). *Albert Einstein and the Cosmic World Order*, New York: Interscience Publishers (John Wiley).

Lanczos, C. (1974). *The Einstein Decade (1905-15)*, London: Elek Science.

Maranto, G. (1985). "Einstein's Brain," *Discover*, Vol.6, No. 5, May, p. 28.

Mayer, J., Holms, J. P. (1996). *Bite-Size Einstein: Quotations on Just About Everything from the Greatest Mind of the Twentieth Century*, New York: St. Martin's Press.

McCausland, I. (1999). "Anomalies in the History of Relativity," *Journal of Scientific Exploration*, 13, 2, 271.

McCrea, W. H. (1973). "The fortunate physicist," *New Scientist*, 13 September, p. 641.

McCrea, W. H. (1982). "Albert Einstein," *New Scientist*, 2 December, p. 582.

Missner, M. (1985). "Why Einstein Became Famous in America," *Social Studies of Science*, Vol. 15, pp. 267-291

Overbye, D. (2000). *Einstein in Love: A Scientific Romance*, New York: Viking.

Pais, A. (1982). *'Subtle is the Lord ...': The Science and the Life of Albert Einstein*. Oxford: Clarendon Press.

Pais, A. (1994). *Einstein Lived Here*. Oxford: Clarendon Press.

Pais, A. (1997). *A Tale of Two Continents: A Physicist's Life in a Turbulent World*, Princeton, NJ: Princeton University Press.

Paterniti, M. (1997). "Driving Mr. Albert: A Trip Across America with Einstein's Brain," New York, *Harper's Magazine*, Vol. 295, No. 1769, October, p. 35.

Paterniti, M. (2000). *Driving Mr. Albert: A Trip Across America with Einstein's Brain*, New York: The Dial Press, Random House, Inc.

Powell, C. S. (2002). *God in the Equation: How Einstein Became the Prophet of the New Religious Era*, New York: The Free Press.

Rowe, A. (1973). "And nobody understands Ein," London: *The Times Higher Educational Supplement*, 14 September, p. 13.

Stachel, J. (1998). *Einstein's Miraculous Year: Five Papers That Changed the Face of Physics*. Princeton, NJ: Princeton University Press.

Thomson, J. (1919). [Chair of] Joint Eclipse Meeting of the Royal Society and the Royal Astronomical Society. London: *The Observatory*, 42, 389.

Ziman, J. (1981). *Puzzles, Problems and Enigmas: Occasional Pieces on The Human Aspects of Science*, Cambridge: Cambridge University Press.

Appendix VII

A Question of Relativity

Professor Herbert Dingle was a long-time critic of the special theory of relativity, who believed for many years that the theory was self-contradictory. Although he was unsuccessful in persuading the scientific world of the inconsistency of the theory, his questions and arguments were not satisfactorily answered during his life. Now, thirty years after his death, the subject is of historical interest. This paper examines two main problems that have contributed to the confusion that still surrounds this issue. The first problem is the fact that some scientists answered Dingle's Question, which is explicitly about the special theory, by invoking the general theory. It is argued that, if there is a valid answer to Dingle's Question, it would have been valid if the same question had been asked in 1905 before the general theory appeared. The second problem is that many scientists have claimed that Dingle's thesis has been refuted by experiment, although experimental results cannot disprove the existence of an internal contradiction. An answer to Dingle's Question is still wanting.

Keywords: Special relativity, Dingle's Question

Introduction

Professor Herbert Dingle (1890-1978) was a distinguished scientist and philosopher of science, who was well known within the scientific community as a long-time critic of Einstein's special theory of relativity. Now, thirty years after his death, his criticisms of the theory are a suitable subject for study by scientists and historians of science.

Dingle's criticisms of special relativity can be divided into two separate phases. At first he criticized the derivation from the theory that is commonly associated with the term clock paradox, according to which a clock that moved at high speed around a closed path, starting and finishing at the position of a stationary clock with which it was initially synchronized, would show a lower reading than the stationary clock when they again came together. An extension of that problem was the twin paradox, in which one of a pair of twins would make a long high-speed journey into space and return to the other twin, and according to the theory the travelling twin would be younger than the stay-at-home twin when they again came together. Dingle's argument was, essentially, that the theory deals only with relative motion so the clocks or twins would both age at the same rate.

The second phase of Dingle's criticism of the special theory started when he realized that he could not disprove the asymmetrical ageing, and he concluded that, if symmetrical and asymmetrical ageing were both predicted by the theory, the special theory must be self-contradictory.

The story of Dingle's criticism of the asymmetrical ageing associated with the clock or twin paradox has been told comprehensively by Chang [1]. Although it is clear from the history of that phase of the controversy that Dingle did not disprove the asymmetrical ageing, he did ask a question about the physical cause of that ageing that Chang believed was not answered. Chang wrote that most of Dingle's opponents were not concerned with physical explanations, and concluded (p. 782) that "What their responses to Dingle came down to was an implicit rejection of his question about the physical cause of the asymmetrical ageing, rather than an answer to it. In my opinion, Dingle's question remains unanswered to this day."

Since Chang's paper was so comprehensive a study of Dingle's arguments about the clock or twin paradox, the present paper does not attempt to deal with that problem any further. Instead, we concentrate on a much more important question that Dingle asked, which in my opinion remains unanswered to this day.

Dingle's QUESTION

The question that Dingle asked was the central theme of his book *Science at the Crossroads* [2], and he claimed that, unless that question could be answered, the special theory failed. This question might be worded very briefly as follows: *Which of two clocks in uniform relative motion does the special theory require to work more slowly?* However, in order to present the story satisfactorily we should consider the question in its extended form, as it is presented on pages 45-46 of his book:

THE QUESTION

> According to the special relativity theory, as expounded by Einstein in his original paper, two similar, regularly-running clocks, A and B, in uniform relative motion, must work at different rates. In mathematical terms, the intervals, dt and dt', which they record between the same two events are related by the Lorentz transformation, according to which dt ≠ dt'. Hence one clock must work steadily at a slower rate than the other. The theory, however, provides no indication of which clock that is, and the question inevitably arises: How is the slower-working clock distinguished? The supposition that the theory merely requires each clock to appear to work more slowly from the point of view of the other is ruled out not only by its many applications and by the fact that the theory would then be useless in practice, but also by Einstein's own examples, of which it is sufficient to cite the one best known and most often claimed to have been indirectly established by experiment, viz. 'Thence' [*i.e.*, from the theory he had just expounded, which takes no account of possible effects of acceleration, gravitation, or any difference at all between the clocks except their state of uniform motion] 'we conclude that a balance-clock at the equator must go more slowly, by a very small amount, than a precisely similar clock situated at one of the poles under otherwise identical conditions.' Applied to this example, the question is: what entitled Einstein to conclude from his theory that the equatorial, and not the polar, clock worked more slowly?

Let us examine the answer to this question that appeared in Ziman's review [3] of Dingle's book. Ziman quoted Dingle's Question, as it is shown above, up to and including the sentence "How is the slower-working clock distinguished?" Immediately after quoting the question, he wrote: "This is a perfectly reasonable question to which science should indeed give an answer." Later in the review he wrote: "To distinguish the slower-working clock (as demanded by Professor Dingle) one must use the theory of *general* relativity, which takes account of accelerations." He gave his answer to the question in the following words: "In fact, the answer to Dingle's 'question' is simple: the fastest working clock between any two events is one that travels between them by free fall."

These quotations make it obvious that Ziman's statement is not a valid answer to Dingle's question. It is clear from the question that the answer is to be provided from the special theory, not the general theory. It is equally clear that Ziman's statement fails to answer the part of the question referring to the polar and equatorial clocks, which Ziman did not include in his quotation of the question.

Since Dingle's question was explicitly asked about the special theory and only the special theory, as it was described in Einstein's 1905 paper, any valid answer to Dingle's question must have the property that it would have been a valid answer if the question had been asked in 1905, just after Einstein's paper was published and before the general theory appeared.

Another scientist, G. F. R. Ellis [4], described Ziman's review as "admirable" but wrote that "Ziman invokes general relativity at a stage when it is not really needed." He went on to say: "In special relativity, just as in general relativity, the answer to Professor Dingle's 'question' is: the fastest working clock between any two events is one that travels between them by free fall."

We see that Ziman wrote that the answer to Dingle's question requires the use of the general theory, whereas Ellis wrote that it does not require it: that is a clear contradiction, on an important property of a theory. If Ziman and Ellis had been interested in the search for the truth of the matter, they might have been expected to try to resolve that contradiction, but there is no record of any further discussion between them on that subject. Although they disagreed on an important property of a theory, they agreed that Dingle was wrong in saying there was a problem. One is reminded of remarks by Cullwick [5] which, although they were written about the earlier debate over the clock paradox, seem appropriate to the later controversy:

> On one thing Professor Dingle's critics are all agreed, that he is wrong. They do not all agree, however, on the nature of his error. Some give arguments which are no more than illustrations of the obvious fact that the reciprocal Lorentz transformation is algebraically consistent; some claim that the problem requires the General Theory of Relativity; and some appear to regard the matter as settled by their knowl-

> edge of four-dimensional space-time. Some argue with patience, while others thinly disguise their irritation.

Ellis's answer [4] does not pass the test of being valid in 1905, before the general theory appeared. Consider the following argument. Suppose that there is a perfectly spherical, non-rotating planet without an atmosphere, and that a clock is carried on a satellite in a circular orbit that is just above the surface of the planet. (For a planet having the same size and mass as our Earth, such an orbit would have a period of about 84 minutes.) Suppose that the orbiting clock passes over a stationary clock on the surface of the planet once every orbit, and that the clock readings are compared at each of these close encounters. According to Einstein's 1905 paper, the reading of the clock that goes round a closed path and returns to the other clock would fall behind the reading of the clock on the surface, but according to Ellis's answer the orbiting clock would work faster than the other because the clock on the satellite is in free fall. Although there were no clocks on satellites in 1905, someone with enough imagination and knowledge could easily have suggested such a thought experiment then to refute that answer to the question.

Like Ziman, Ellis did not show how his answer applied to the polar and equatorial clocks. Dingle [6] replied to him and Ziman as follows: "Professor J. Ziman and Mr. G. F. R. Ellis seem not to have read my 'question', let alone answered it, though Ziman quotes it correctly. Neither of the events need be at either of the clocks concerned, so the statement, 'the fastest working clock between any two events is one that travels between them by free fall', is futile."

Another reviewer of Dingle's book also used the general theory to try to rescue the special theory from its problems. Referring to the polar and equatorial clocks, Stadlen [7] wrote as follows:

> But the relative motion involved in this case, being circular, is non-uniform. I submit, therefore, that Einstein was wrong in saying that his prediction followed from the special theory, which deals only with the effects of uniform motion. This is not to say that the prediction was invalid. For Einstein was, intuitively, anticipating his later general theory, according to which the equatorial clock runs slower because of the centripetal force exerted upon it.

The fact that the predicted slowing follows from the general theory does not make Einstein's prediction *from the special theory* valid; it is a well known fact of logic that the truth of the conclusion of an argument does not guarantee the validity of the argument. If Einstein's prediction does not follow from the special theory, then his inclusion of that prediction in his 1905 paper was irrational and, therefore, not valid. Also, the suggestion that Einstein was so easily able to anticipate his general theory, which took him about another decade to develop, is highly unconvincing.

Obviously Stadlen's statement could not have been a valid answer to Dingle's question if it had been asked in 1905. Also, anyone who claims that Einstein's statement about the polar and equatorial clocks is wrong, as Stadlen does, ought to be able to state categorically at exactly what point in Einstein's reasoning he went wrong.

The Problem of Inconsistency

One of the biggest problems in dealing with Dingle's criticisms of the special theory is that many scientists seem to have very poor knowledge of simple logic. Because of their poor knowledge, they do not fully understand the properties of a theory that has an internal inconsistency or contradiction, and this causes them to use irrational arguments in trying to refute Dingle's claim that the special theory is internally inconsistent.

It is a well-known fact of simple logic that, if a theory has an internal contradiction or inconsistency, it is possible to derive from that theory any result that we wish. As Popper [8] puts it: "But the importance of the requirement of consistency will be appreciated if one realizes that a self-contradictory system in uninformative. It is so because any conclusion we please can be derived from it."

That fact has enormous significance in assessing the responses of scientists to Dingle's claim that there is an internal contradiction. If a theory is self-contradictory, then it is possible to derive results from the theory to match the results of any physical experiments that we choose. That means that the claim that a theory is self-contradictory cannot be refuted by saying that the theory matches various experimental results, since a self-contradictory theory could match all possible results.

The question of whether the special theory is self-contradictory can be answered only by studying the theory. Imagine that, in 1905, a suitably-qualified scientist had been confined to a sealed room with a copy of Einstein's paper and asked to find out whether the theory described in the paper was self-consistent. He might not have been able to prove that the theory was self-consistent, but it is quite certain that, if he had found a contradiction, nothing that ever happened outside that room, either the performance of experiments or the publication of the general theory, could remove that contradiction.

In short, in assessing the merits of Herbert Dingle's claim that the special theory of relativity is self-contradictory, all experimental results whatever are completely irrelevant.

Consider, for example, another review of *Science at the Crossroads* that appeared in *Nature* in the form of an unsigned editorial article [9]. The last two sentences of the review, which are presented triumphantly as if they were a refutation of Dingle's thesis, read as follows:

> And is there any hope that he will now be satisfied with the demonstration that moving clocks run at different speeds from clocks at rest which has been provided in the past few months by the experiments in which Hafele and Keating have flown caesium clocks in different directions around the world (*Science*, **177**, 166; 1972, see also *Nature*, **238**, 244; 1972)? It will be sad to see the clock paradox disappear, but this work is the last nail in the coffin.

It would seem reasonable to expect that a scientist who was qualified to write a *Nature* editorial would know enough elementary logic to realize that the experiment did not refute Dingle's thesis, but unfortunately that condition is not satisfied. Also, the writer seems not to have noticed Dingle's statement that he had for years held an open mind on the subject of asymmetrical ageing, or his statement that the scientific issue was not what is normally associated with the expression "clock paradox," both of which appear in the Preface of *Science at the Crossroads.*

Consider also the following quotation from another unsigned *Nature* editorial that is reproduced as an appendix in Dingle's book [2, p. 226]:

> But in circumstances like these, where a theory is lent conviction by the sheer breadth of its agreement with experiment, it would seem incumbent on those who would overthrow it to produce not merely a contradiction but a constructive alternative.

Not only does the writer of this strange argument not appear to understand that the agreement with experiment does not disprove the possible existence of a contradiction, but he also refuses to accept that a contradiction—an obviously fatal flaw in a theory—is sufficient to require scientists to discard the theory unless the person who demonstrates the contradiction also presents an alternative theory. An argument like this was criticized by Nordenson [10], who called it a grotesque argument and made the following comment:

> It is as if a judge, who has established that a person, accused of murder, has brought out evidence of his innocence, would declare: 'It may well be that you have proved your innocence but I cannot release you unless you find who the murderer is.'

It is interesting that Dingle, even though he reproduced the above-mentioned *Nature* editorial as an appendix in his book, did not make the obvious criticism that "the sheer breadth of its agreement with experiment" did not refute his claim of a contradiction. I think it needs to be said that Dingle himself did not seem to realize fully the irrelevance of experimental evidence. In various places he discussed the experimental evidence for the theory, saying for example that much of the evidence in its favour depended on circular arguments because the velocities of elementary particles were inferred from electromagnetic theory rather than being found by measuring the distance actually travelled by an individual particle in a given time. Perhaps he should simply have pointed out that all experimental evidence is irrelevant in attempting to refute a claim that the theory is self-contradictory.

The fact that any statement we wish can be derived from a self-contradictory theory is also highly significant in assessing the debates about the twin paradox that have gone on for many years. For example, suppose that one scientist argues that, according to the special theory, an astronaut who makes a long round-trip journey at very high speed would age less than his sibling who stayed at home, and suppose that another scientist argues that the two would age equally. It is natural for each scientist to believe that, since his own argument appears to be correct, the other scientist's argument must be wrong. But it is not necessarily so: if the special theory was self-contradictory, both arguments might be correctly based on the postulates of the theory. In other words, the possibility that the theory is self-contradictory could account for the long inconclusive series of debates on the twin paradox. If the theory was self-contradictory it would also make it impossible to disprove the theory by experiment, because supporters of the theory could show that the theory matched any experimental results whatever.

Discussion

As mentioned above, Ziman wrote that Dingle's Question is "a perfectly reasonable question to which science should indeed give an answer." I believe it is obvious that the question has not been satisfactorily answered, but it is not so obvious who should answer it. Ziman did not say who should answer it: he asked "what do we mean by 'science' in this context?" and went on to say that "physics has no Pope with authority to proclaim doctrine." Of course he is correct in saying that, and it is obvious that science, as an abstract idea, cannot give an answer; only scientists can answer the question.

I believe that future historians of science will be very puzzled by the fact that, in spite of the ineptitude of the published attempts to answer Dingle's Question and his other arguments, the scientific world remains almost unanimous to this day in its belief that Dingle was all wrong and his opponents all right. Although I have quoted in this paper only a few attempts to answer Dingle's arguments, I have shown elsewhere [11,12] that several of Dingle's opponents contradicted one another in their attempts to show that there is no contradiction in the special theory.

I suggest that scientists need to answer Dingle's Question. Since the question was explicitly about the special theory of relativity, I suggest that the answer to the question should have the following properties: it should provide a clear criterion to distinguish which of two clocks in relative motion the special theory requires to work more slowly, the applicability of the criterion to the case of the polar and equatorial clocks should be clear, and the answer should not depend in any way on the general theory of relativity. The world has been waiting more than thirty-five years for Herbert Dingle's perfectly reasonable question to be answered.

References

[1] H. Chang, "A Misunderstood Rebellion: The Twin-Paradox Controversy and Herbert Dingle's Vision of Science," *Studies in History and Philosophy of Science* **24**, (1993), 741-790.

[2] H. Dingle, *Science at the Crossroads*, Martin Brian & O'Keeffe, London, (1972).

[3] J. Ziman, "Science in an Eccentric Mirror," *Nature* **241**, (12 January 1973), 143-4.

[4] G. F. R. Ellis, "Special Relativity Again," *Nature* **242**, (9 March 1973), 143.

[5] E. G. Cullwick, "The Riddle of Relativity," *Bulletin of the Institute of Physics* **10**, (March 1959), 52-7.

[6] H. Dingle, "Dingle's Question," *Nature* **242**, (6 April 1973), 423.

[7] G. Stadlen, "Dingle's Challenge," *The Listener* **88**, (28 September 1972), 411-2.

[8] K. R. Popper, *The Logic of Scientific Discovery*, Harper Torchbooks, New York, Second Edition. (1968), p. 92.

[9] Anonymous, "Dingle's Answer," *Nature* **239**, (29 September 1972), 242.

[10] H. Nordenson, *Relativity Time and Reality*, George Allen and Unwin Ltd., London, (1969), p. 102n.

[11] I. McCausland, "Science on the defensive," *Canadian Electrical Engineering Journal* **5**, No. 2, (1980), 3-4.

[12] I. McCausland, "Problems in Special Relativity," *Wireless World* **89**, No. 1573, (October 1983), 63-65.

Bibliography

Abe, F. *et al.* 1991. In *Physical Review Letters* **67**, No. 19, 4 November, 2609.

Al-Khalili, J. 2011. If this is right, I will eat my boxer shorts on TV. *The Times*, 24 September.

Anon. 1919. Joint Eclipse Meeting of the Royal Society and the Royal Astronomical Society. *The Observatory*, **42**, No. 545, November, 389-398.

_____. 1967. Don't Bring Back the Ether. *Nature*, **216,** October 14, 113-4

_____. 1972. Dingle's Answer. *Nature*, **239**, September 29, 242.

_____. 1983 (Identified as "J. P. S.") Relativity Theory for Everyman. *Physics in Canada*, **39**, No. 2, March, 52-53.

Armstrong, H. L. 1973. Freely Falling Clocks, *Nature*, **244**, July 6, 26.

Armytage, W. H. G. 1957. *Sir Richard Gregory: His Life and Work*, London, Macmillan.

Banks, D. L. 1996. A Conversation with I. J. Good. *Statistical Science*, **11**, No. 1, 1-19.

Barnett, L. 1949. *The Universe and Dr. Einstein*. Victor Gollancz, London.

Barrow, J.D. 1988. *The World within the World*. Clarendon Press, Oxford.

Bauer, H. H. 1984. *Beyond Velikovsky: The History of a Public Controversy*. University of Illinois Press.

__________. 2006. Editorial. *Journal of Scientific Exploration*. **20**, No. 4, 511-4.

__________. 2006. Editorial. *Journal of Scientific Exploration*. **21**, No. 3, 437-441..

Beckmann, P. 1990. Review of *The Relativity Question*. *Galilean Electrodynamics*, **1**, No. 1, January, 12.

Berezinsky, V. 1973. How a theoretical physicist works. In *A Random Walk in Science*, compiled by R. L. Weber. The Institute of Physics.

Bergson, H. 1965. *Duration and Simultaneity: with reference to Einstein's Theory*. Translated by Leon Jacobson, with an Introduction by Herbert Dingle. The Bobbs-Merrill Company, Inc., Indianapolis.

Berry, M. and J. F. Nye. 2006. John Michael Ziman. *Biographical Memoirs of Fellows of the Royal Society*, **52**, 479-491.

Bodanis, D. 2004. Inventor of the Modern World. *Discover*, September, 32-35.

Bondi, H. 1964. *Relativity and Common Sense*. Anchor Books, Doubleday & Company, Inc., New York.

Born, Max 1968. *My Life & My Views*. Charles Scribner's Sons, New York, 183.

Bowersock, G. W. 1984. The Art of the Footnote. *The American Scholar*, **53**, 54-62.

Brockman, J. 2006. (Editor) *My Einstein: Essays by Twenty-four of the World's Leading Thinkers on the Man, His Work, and His Legacy*, Pantheon Books, New York.

Bronowski, J. 1961. Dr Bronowski replies to Professor Dingle. *New Scientist*, 31 August, 542.

Brown, G. Burniston 1967. What is Wrong with Relativity? *Bulletin of the Institute of Physics*, March, 71-77.

Buller, A. H. R. 1923. Relativity. *Punch*, **165**, December 19, 591. (Note: Although the item appeared anonymously, Buller's authorship is well established (Goldsborough, 2004).)

Burnell, S. J. 1978. Dingle and the Scientists.*The Friend*, October 27, 1340.

Calaprice, A. 1996. (Editor) *The Quotable Einstein*. Princeton University Press.

___________. 2000. (Editor) *The Expanded Quotable Einstein*. Princeton University Press.

___________. 2002. (Editor) *Dear Professor Einstein: Albert Einstein's Letters to and from Children*. Prometheus Books, Amherst, New York.

___________. 2005a. (Editor) *The New Quotable Einstein*. Princeton University Press.

___________. 2005b. *The Einstein almanac*. Johns Hopkins University Press.

___________. 2011. (Editor) *The Ultimate Quotable Einstein*. Princeton University Press.

Chang, H. 1993. A Misunderstood Rebellion: The Twin-Paradox Controversy and Herbert Dingle's Vision of Science. *Studies in History and Philosophy of Science*, **24**, 741-790.

Chappell, J. 1980. Letter to the Editor. *Speculations in Science and Technology*, **3**, No. 4, 488-495.

Coleman, J. A. 1954. *Relativity for the Layman: A Simplified Account of the History, Theory, and Proofs of Relativity*. William-Frederick Press, New York.

____________. 1959. *Relativity for the Layman: A Simplified Account of the History, Theory, and Proofs of Relativity*. Penguin Books.

____________. 1969. *Relativity for the Layman: A Simplified Account of the History, Theory, and Proofs of Relativity*. Revised Edition, described on back cover as Second Edition. Penguin Books.

____________. 1996. *Tales of a Scientist*. Priority Press, East Longmeadow, MA.

Cullwick, E. G. 1959. The Riddle of Relativity. *Bulletin of the Institute of Physics*, **10**, 52-57 (March).

___________. 1976. Relativity and the ether. *Electronics & Power*, January, 40.

___________. 1981. Einstein and Special Relativity: Some Inconsistencies in His Electrodynamics. *Brit., J. Phil. Sc.*, **32**, 167-176.

Cunningham, E. 1914. *The Principle of Relativity*. Cambridge University Press.

_____________. 1921. *Relativity and the Electron Theory*. Longmans, London.

Davies, P. 1980. Why pick on Einstein? *New Scientist*, 7 August 1980, 463-5.

______. 1983. *God and the New Physics*, J. M. Dent & Sons, London.

______. 1995. *About Time: Einstein's Unfinished Revolution*, Simon & Schuster.

Dingle, H. 1922. *Relativity for All*. Methuen.

________. 1928. (Identified as "H.D.") The Understanding of Relativity. *Nature*, **122**, November 3, 673-5.

________. 1931. *Science and Human Experience*, Williams & Norgate Ltd.

________. 1939. The Interpretation of the Michelson-Morley and Kennedy-Thorndike Experiments. *Philosophical Magazine*, **27**, January-June 1939, 693-702.

_________. 1940. *The Special Theory of Relativity*. Methuen.

_________. 1952. *The Scientific Adventure: Essays in the History and Philosophy of Science*. Pitman, London.

_________. 1956. Relativity and Space Travel. *Nature*, **177**, April 28, 782-4. Followed by McCrea's reply (784-5) and Dingle's further response (785).

_________. 1960. Relativity and Electromagnetism: An Epistemological Appraisal. *Philosophy of Science*, **27**, No. 3, 233-253.

_________. 1961a. *The Special Theory of Relativity*. Methuen, Fourth edition..

_________. 1961b. An improved method of youth preservation? *New Scientist*, 24 August, 455-457.

_________. 1962. Special Theory of Relativity. *Nature*, **195**, September 8, 985-6.

_________. 1969. Definitions and Realities–the Modern Delusion. *The Listener*, **82**, July 3, 16-17.

__________. 1972. *Science at the Crossroads*, Martin Brian & O'Keeffe, London.

__________. 1973a. Dingle's Question. *Nature*, **242**, April 6, 423.

__________. 1973b. Dingle's Answer. *Nature*, **243**, June 8, 366.

__________. 1975a. Integrity in science. *Nature*, **255**, June 12, 519-20.

__________. 1975b. Integrity in science. *Nature*, **256**, July 17, 162.

__________. 1977a. Letter To the Editor. *British Journal for the History of Science*, **10**, Part1 (March) 94.

_________. 1977b. The twin paradox. *Nature*, **269**, 22 September, 284.

__________. 1979. Time in Philosophy and in Physics. *Philosophy*, **54**, 99-104.

__________. 1980. The "twins" paradox of relativity. *Wireless World*. **86**, No. 1537, October 1980, 54-56.

Dirac, P. A. M. 1930. *Principles of Quantum Mechanics*. Clarendon Press, Oxford.

____________. 1963. The Evolution of the Physicist's Picture of Nature. *Scientific American*, **208**, No. 5, May, 45-53.

____________. 1966. *Lectures on Quantum Field Theory*. Yeshiva University.

____________. 1980. Why We Believe in the Einstein Theory. In *Symmetries in Science*, edited by B. Gruber and R. S. Millman, pp. 1-11. Plenum Press.

Dyson, F. J. 1968. Interstellar Transport, *Physics Today*, October, 41-45.

Eddington, A. S. 1959. *New Pathways in Science*. University of Michigan Press, 211.

Einstein, A., and L. Infeld. 1949. On the Motion of Particles in General Relativity Theory. *Canadian Journal of Mathematics*. **1**, 209-241.

Einstein, A. 1989. *The Collected Papers of Albert Einstein*, **2** (The Swiss Years: Writings, 1900-1909), Princeton University Press.

Ellis, G. F. R. 1973. Special Relativity Again. *Nature*, **242**, March 9, 143.

Evans, C. 1979. Interview with I. J. Good, *Omni*, **1**, No. 4, January, 71-73, 117-121.

Feyerabend, P., 1981. How to Defend Society Against Science. In I. Hacking (Ed.) *Scientific Revolutions*, 156-167. Oxford University Press.

Frisch, O. 1979. The Life and Times of Albert Einstein. *New Scientist*, **81**, 753-755.

Gaughan, J. A. 1986. *Alfred O'Rahilly, I: Academic.* Kingdom Books, Mount Merrion, County Dublin.

Gift, S. J. G. 2006. The Relative Motion of the Earth and the Ether Detected. *Journal of Scientific Exploration.* **20**, 201-214.

Glaisher, J. W. L. (Ed.) 1894. *The Collected mathematical Papers of Henry John Stephen Smith*, Vol. I. Clarendon Press, Oxford.

Goldberg, S. 1984. *Understanding Relativity.* Birkhäuser, Boston.

Goldsborough, G. 2004. Reginald Buller: The Poet-Scientist of Mushroom City. *Manitoba History: The Journal of the Manitoba Historical Society.* No. 47, Spring/Summer. (Available online at the website of The Manitoba Historical Society.)

Good, I. J. 1962. (General Editor.) *The Scientist Speculates: An Anthology of Partly-Baked Ideas.* Heinemann, London.

_________. 1998. "Establishment Thinking" A reply to Chappell. *Apeiron*, **5**, Nos. 3-4 (July-October 1998), 242-244.

_________. 2008. Review of *Old Physics for New* by Thomas E. Phipps, Jr., *Physics Essays*, **21**, 313-4.

Grant, J. 2006. *Discarded Science.* Artists and Photographers' Press Ltd., Wisley, Surrey.

Haldane, J. S. (Ed.) 1928. *The Collected Scientific Papers of John James Waterston.* Oliver and Boyd, Edinburgh.

Hardy, G. H. 1940. *A Mathematician's Apology.* Cambridge University Press.

Harvey A. and E. Schucking, 2005. A Small Puzzle from 1905. *Physics Today*, **58**, issue 3, March, 34-36.

Hawking, S. 1988. *A Brief History of Time.* Bantam Press.

__________. (Ed.) 2002a. "Selections from *The Principle of Relativity.*" Running Press, Philadelphia and London.

__________. (Ed.) 2002b. *On The Shoulders of Giants; The Great Works of Physics and Astronomy.* Running Press, Philadelphia and London.

Hayes, P. 2010. Popper's response to Dingle on special relativity and the problem of the observer. *Studies in History and Philosophy of Modern Physics*, **41**, Issue 4, November, 354-361.

Helliwell, T. M. 1966. *Introduction to Special Relativity.* Allyn and Bacon.

Hentschel, K. 1992. Einstein's Attitude Towards Experiments: Testing Relativity Theory 1907-1927. *Stud. Hist. Phil. Sci.*, **23**, No. 4, 593-624.

Holton, G. 1960. On the Origins of the Special Theory of Relativity. *American Journal of Physics*, **28**, 627-636.

________. 1992. Book Review. *Foundations of Physics*, **22**, No. 12, 1517-1520.

________. 1993. *Science and Anti-Science.* Harvard University Press.

Huxley, T. H. 1894. The Coming of Age of "The Origin of Species." In *Darwinia* (Collected Essays, Volume II.) Macmillan & Co., London.

Jaki, S. 1988. *The Absolute beneath the Relative and Other Essays.* University Press of America.

_____. 1989. *God and the Cosmologists.* Scottish Academic Press.

Jammer, M. 2004. The Strange Story of the Concept which Inaugurated Modern Theoretical Physics. *Foundations of Physics*, **34**, No. 11, 1617-1641.

Janssen, M. 2006. 2005: The centenary of Einstein's *annus mirabilis. Studies in History and Philosophy of Modern Physics*, **37**, 1-4.

Jeans, J. 1931. *The Mysterious Universe*. Second edition, Cambridge University Press.

Johnson, P. 1991. *A History of the Modern World: From 1917 to the 1990s*. Weidenfeld and Nicolson, Revised edition.

Johnston, R. 1973. The case of Dingle versus Einstein. *The Irish Times*, January 10.

Judson, H. F. 1980. The Rage to Know: Science is the true modern art. *The Atlantic Monthly*, **245**, No. 4, April, 112-117.

______. 2004. *The Great Betrayal: Fraud in Science*, 406-7. Harcourt, Inc..

Kaufmann, W. 1968, *Philosophic Classics, Volume II: Bacon to Kant*. Prentice-Hall, Second Edition.

Lanczos, C. 1965. *Albert Einstein and the Cosmic World Order*. Interscience.

Landsberg, P. 1995. Much about time in little space. (Review of *About Time: Einstein's Unfinished Revolution*, by Paul Davies.) *Nature* **375**, 15 June, 548.

Levenson, T. 2003. *Einstein in Berlin*. Bantam Books, New York.

Levin, B. 1961. A plain man's guide to the theory that Einstein was wrong. *The Times* (London) 21 December.

Lockyer, T. M. 1928. *Life and work of Sir Norman Lockyer*. Macmillan.

Loeb, L. B. and A. S. Adams. 1933. *The Development of Physical Thought: A Survey Course of Modern Physics*. Wiley, New York.

Lorentz, H. A., A. Einstein, and H. Minkowski, 1913. *Das Relativitätsprinzip*. Teubner. Edited by Otto Blumenthal.

Lorentz, H. A., A. Einstein, H. Minkowski and H. Weyl. 1923. *The Principle of Relativity*. Methuen, republished by Dover in 1952.

Macbeath, A. 1952. *A Plea for Heretics*. Evening discourse at The Annual Meeting of the British Association for the Advancement of Science in Belfast. The Queen's University of Belfast.

Maddox, J. 1975. Integrity in science. *Nature*, **255**, 12 June, 520.

________. 1986. More tests of special relativity. *Nature*, **319**, 13 February, 533.

________. 1996. By any other name... *Nature*, **383**, 5 September, 17.

Marder, L. 1971. *Time and the Space Traveller*. George Allen & Unwin Ltd.

Maudlin, T. 1994. *Quantum non-locality and relativity: metaphysical intimations of modern physics*. Blackwell.

May, D. 2001. *Critical Times: The History of The Times Literary Supplement*. HarperCollins Publishers, 84.

McCartney, M. and A. Whitaker (Eds.) 2003. *Physicists of Ireland: Passion and Precision*, Institute of Physics Publishing, Bristol and Philadelphia.

McCausland, I. 1972. On the Special Theory of Relativity. *Proc. IEE*, **119**, pp. 1766-7 (Also *Proc. IEE*, **120**, 1973, 527-528).

___________. 1974. An Analogy Between the Lorentz Transformation and the Transmission-Line Equations. *Proc. IEEE*, **62**, 1398-1399.

___________. 1980a Science on the Defensive. *Canadian Electrical Engineering Journal*, **5**, No. 2, 3-4.

___________. 1980b. Binary Stars and the Velocity of Light. *Speculations in Science and Technology*, **3**, No. 4, October, 509-511.

___________. 1983a. Einstein. *Times Literary Supplement*, April 22, 404.

___________. 1983b. Problems in Special Relativity. *Wireless World*, October, 63-65.

___________. 1984. Einstein and Special Relativity: Who Wrote the Added Footnotes? *Brit., J. Phil. Sc.*, **35**, 60-1.

___________. 1988. *The Relativity Question*. Privately published. Available online at http://www.worldsci.org/people/Ian_McCausland.

___________. 1990. An Inconsistency in Special Relativity. *Physics Essays*, **3**, No. 2, 176-7.

___________. 1992. On Faraday's Ray-vibrations and the Postulates on the Velocity of Light. *Physics Essays*, **5**, No. 3, 357-361.

___________. 1999. Anomalies in the History of Relativity. *Journal of Scientific Exploration*, **13**, No. 2, , 271-290.

___________. 2003. The Einstein Mystique. *Journal of Scientific Exploration*, **17**, No. 4, 715-730.

___________. 2005. The Persistent Problem of Special Relativity. *Physics Essays*, **18**, No. 4, 530-532.

___________. 2008. A Question of Relativity. *Apeiron*, **15**, April, pp. 545-557. Available online at: http://redshift.vif.com/journal_archives.htm

McCrea, W. H. 1935. *Relativity Physics*. Methuen, London.

____________. 1951. The Clock Paradox in Relativity Theory. *Nature*, **167**, April 28, 680.

____________. 1957. Relativistic Ageing. *Nature*, **179**, May 4, 909-910.

____________. 1969. Definitions and Realities. *The Listener*, 4 September.

McDougall, W. 1927. *Character and the Conduct of Life: Practical Psychology for Everyman*. Methuen, London, Second edition, xii.

Mill, J. S. 1874. *On Liberty*. Longmans, Green, Reader and Dyer, Fifth edition.

Moffatt, J. W. 2010. *Einstein Wrote Back: My Life in Physics*. Thomas Allen Publishers, Toronto.

Moore, M. G. 1959. Review of Coleman's *Relativity for the Layman*. *Science*, **129**, 16 January, 144.

Mulkay, M. J. 1972. *The Social Process of Innovation: A study in the sociology of science*. Macmillan, 19.

Nordenson, H. 1969. *Relativity Time and Reality: A critical investigation of the Einstein Theory of Relativity from a logical point of view*. George Allen and Unwin Ltd., London.

Ohanian, H. C. 2008. *Einstein's Mistakes: The human failings of genius*. Norton.

O'Rahilly, A. 1938. *Electromagnetics*. Cork University Press. Later published in two volumes as *Electromagnetic Theory*. Dover, 1965.

Osborne, M. S. 2006. Response to "The Relative Motion of the Earth and the Ether Detected." *Journal of Scientific Exploration*. **20**, 597.

Pais, A. 1982."*Subtle is the Lord ...": The Science and the Life of Albert Einstein*. Oxford University Press.

_______. 1994. *Einstein Lived Here*. Oxford University Press, 33.

Park, R. L. 2008. *Superstition: Belief in the Age of Science*. Princeton University Press.

Pausch, R., with J. Zaslow. 2008. *The Last Lecture*. Hyperion.

Phipps, T. E. Jr. 2006. *Old Physics for New*. Montreal: Apeiron.

Pippard, B. 1983. To the heart of matter. *Times Literary Supplement*, April 1, 315.

Planck, Max 1949. *Scientific Autobiography and Other Papers*.. Philosophical Library, New York. 33-34.

Polanyi, M. 1958. *Personal Knowledge: Towards a Post-Critical Philosophy*. Routledge & Kegan Paul.

Popper, K. 1968. *The Logic of Scientific Discovery*. Second Edition. Harper & Row.

________. 1971. Einstein's Error. *The Listener*, 22 July, 1971, p. 115.

________. 1982. *Quantum Theory and the Schism in Physics*. Rowan and Littlefield.

Powell, C. 2002. *God in the Equation: How Einstein Became the Prophet of the New Religious Era*. The Free Press.

_________ 2006. My Three Einsteins. In Brockman (2006), pp. 95-107.

Pyenson, L. 1976. Review of *Science at the Crossroads*. *British Journal for the History of Science*, **9**, 336-7.

Reich, E. S. 2011. Speedy neutrinos challenge physicists. *Nature*, **477**, 29 September, 520.

Roach, M. 2010. *Packing for Mars*. Norton.

Rothman, Tony. 2003. *Everything's Relative: and other fables from science and technology*. Wiley.

Rowan-Robinson, M. 1969a. Definitions and Realities. *The Listener*, 7 August, 189.

__________________. 1969b. Definitions and Realities. *The Listener*, 18 September, 380.

__________________. 1978. Rebel without a cause. *Nature*, **276**, 9 December, 150-1.

__________________. 1996. *Cosmology*. Clarendon Press, Oxford, Third Edition, 143.

Roxburgh, I. 1972. Is special relativity right or wrong? *New Scientist*, 55, 28 September, 602.

Samuel H. and H. Dingle. 1961. *A Threefold Cord: Philosophy, Science, Religion*. George Allen & Unwin Ltd., London.

Sarkar, S. 2007. *Doubting Darwin? Creationist Designs on Evolution*. Blackwell Publishing,

Sarty, G. E. 1998. The Twin Paradox? *Journal of the Royal Astronomical Society of Canada*, **92**, 93-5.

Silver, B. L. 1998. *The Ascent of Science*. Oxford University Press, 431.

Smolin, L. 2006.*The Trouble with Physics*. Houghton Mifflin.

Sperling, N. 2002. *What Your Astronomy Textbook Won't Tell You*. Everything in the Universe, San Mateo, California.

Stachel, J. 1998. *Einstein's Miraculous Year*. Princeton University Press, 159-160.

________. 1995. History of Relativity. In *Twentieth Century Physics*, Volume I, Edited by L. M. Brown, A. Pais, and B. Pippard. Chapter 4, pp. 249-356. Institute of Physics Publishing, and American Institute of Physics Press.

________. 2002. *Einstein from 'B' to 'Z'*. Birkhäuser.

Stadlen, G. 1972. Dingle's Challenge. *The Listener*, 28 September, 411-2

Stanley, M. 2007. *Practical Mystic: Religion, Science, and A. S. Eddington*. The University of Chicago Press.

Taylor, J. 1971a. A Warning to Scientists. (Review of Arthur Koestler's *The Case of the Midwife Toad.*) *The Listener*, **86**, 7 October, 483-4.

_______. 1971b. Views. *The Listener*, **86**, 11 November, 642-3.

Thomson, G. 1955. *The Foreseeable Future*. Cambridge University Press.

Velikovsky, I. 1983. *Stargazers and Gravediggers*. William Morrow and Company, Inc., New York.

Waldron, R. A. 1977. *The Wave and Ballistic Theories of Light–A Critical Review*. Frederick Muller Limited, London.

Waller, J. 2002a. *Fabulous Science: Fact and Fiction in the History of Scientific Discovery*. Oxford University Press.

______. 2002b. *Einstein's Luck: The Truth Behind Some of the Greatest Scientific Discoveries*. Oxford University Press.

White, J. L. 1973. Review of *Science at the Crossroads*. *Journal of the British Astronomical Association*. **84**, 70-71.

Whitrow, G. J. (Ed.) 1967. *Einstein: The Man and his Achievement*. Dover.

__________. 1975. Review of *Science at the Crossroads*. *British Journal for the Philosophy of Science*, 26, pp. 358-362.

___________. 1979. Herbert Dingle. *Nature*, **277**, 15 February, 584-5.

___________. 1980. Herbert Dingle, *Quarterly Journal of the Royal Astronomical Society* **21**, 333-337.

Whittaker E. T. 1953. *A History of the Theories of Aether and Electricity, Volume Two, The Modern Theories 1900-1926*. Nelson.

Wilkie, T. D. B. 1977. The Twin Paradox Revisited. *Nature*, **268**, 28 July, 295-6.

____________. 1981. Twins paradox of Relativity. *Wireless World*, June, 47-8.

Will, C. 1986. *Was Einstein Right? putting general relativity to the test*. Basic Books.

Williams, I. 2003. Sir William McCrea 1904-1999. In M. McCartney and A. Whitaker (Eds.), *Physicists of Ireland: Passion and Precision*. Institute of Physics Publishing, Bristol and Philadelphia.

Zapffe, C. A. 1982. *A Reminder on* $E = mc^2$, ... CAZlab.

Ziman, J. 1968. *Public Knowledge: The social dimension of science*. Cambridge University Press.

_______. 1973. Science in an Eccentric Mirror. *Nature*, **241**, January 12, 143-4.

_______. 1977. A plutonium economy. *The Times*, April 27.

_______. 1984. *An Introduction to Science Studies: The Philosophical and Social Aspects of Science and Technology*. Cambridge University Press, 49.

Zweig, Hans J. 2005. *Relativity Unraveled: A Question of Time*. Wiesbaden: Edition Steinherz.

Index

Made in the USA
Monee, IL
04 November 2026

67374728R00144